DYNAMIC COMMUNICATION FOR ENGINEERS

Richard H. McCuen, Peggy A. Johnson and Cynthia Davis

Published by the
American Society of Civil Engineers
345 East 47th Street
New York, New York 10017-2398

ABSTRACT

Dynamic Communication for Engineers provides concise coverage of a broad array of topics in professional/technical communication including oral, written, quantitative, kinetic, and ethical. The "how-to" approach places emphasis on the rules for successful communication, with important do's and don'ts highlighted for ease in application. It will help the engineering student make the transition into the workplace and the young professional complete communication responsibilities more efficiently. Topics covered in this book include procrastination, technical writing style, communication of technical data and statistics, oral presentations and graphics, business correspondence, resumes, job interviews, and nonverbal communication.

Library of Congress Cataloging-in-Publication Data

McCuen, Richard H.
    Dynamic communication for engineers / by Richard H. McCuen, Peggy A. Johnson, Cynthia Davies.
      p.cm.
      ISBN 0-87262-856-6
      1. Communication in engineering. 2. Engineering — Management. I. Johnson, Peggy A. II. Davies, Cynthia. III. Title.
TA158.5.M37 1993
620'0014—dc20   93-9395
                           CIP

TABLE OF CONTENTS

PREFACE

We live in an information age. Information is the prime commodity in today's business world just as manufactured goods were in the nineteenth century and agricultural products were in the eighteenth century. In order to exchange, barter, receive, send, and control information, it is necessary to communicate efficiently and effectively. Technical professionals who cannot or are unwilling to communicate operate under a real disadvantage when it comes to career advancement. It is not enough to invent an innovative process or piece of equipment; today's engineer must be able to communicate his or her work to peers, colleagues, and especially to management. One's ability to communicate directly affects project funding, promotions, and recognition.

Dictionary definitions of communication are usually limited to the act of transmitting ideas or messages. In professional life in a technological age, we must not limit the idea of communication to just the transmission of information. Communication should be evaluated using the criteria of accuracy, efficiency, and effectiveness, all of which influence the success of communications in meeting the requirements of professional life.

In addition to the professionally oriented definition of communication, it is necessary and important to place communication in a broader perspective. Preparing effective transparencies for an oral presentation or knowing the proper formats for headings in written reports are important. But to meet the constraints of effectiveness, efficiency, and accuracy in communication, one has to know how to get started, how to maintain a motivated attitude, how to organize communication responsibilities, and how to incorporate values into professional communications. Developing an efficient writing style and using the latest technology can improve both one's communication ability and effectiveness.

Professional communication goes beyond the dictionary definition because of the important criteria of accuracy, efficiency, and effectiveness and because of the broader perspective required in application within professional life. Communication is a multifaceted concept. It is psychological as well as technical. Very often, the hardest part of communication is getting started; for this reason, we have included Chapter 2 on procrastination. Writing style, the subject of Chapter 3, has psychological as well as technical roots. In the information age, it is not sufficient for the professional to just present a tabular or graphical summary of data. The proper interpretation of data, which is the topic of Chapter 4, is the key to effective communication. Communication is not value neutral; there are important ethical considerations (Chapter 5). Communication involves the written word (Chapters 6 and 9), the spoken word (Chapters 7 and 8), dress and appearance (Chapter 12), and the body language (Chapter 12). The resume (Chapter 10) and the job interview (Chapter 11) are two specific examples of communication that have very important, and obvious, professional implications.

We hope this book will be a convenient guide for both young professionals and students. The style is deliberately informal and conversational, so it will to be readable and easily understood. Conceivably, the student could read through it in one or two sittings, return to it to reflect on the do's and don'ts and practice the exercises, and then keep it on his or her desk as a handy reference tool.

It is just as important to state what the book is not. The book is not intended to be a stand-alone technical writing text for a writing course. Instead, it is a cross between a reference book and a planning guide. For the college students nearing graduation, the book will help them with their job search—resume writing, job interviewing, and dressing for success. For the practicing professional, the book is a reference for many details about writing and speaking. Given this intent, we want to point out that it could also be used as part of a writing course, but supplementary resources would be required.

The book reflects the combined experience and ideas of three professional technical writers: two engineers and one English teacher. It should be of interest to all students who need to practice professional writing skills and strategies. Our intention is to assist the student at every step of the communication process: from writing an effective sentence to analyzing statistical data to landing a job.

We wish to acknowledge the contribution of Ms. Rochelle Leathers to the completion of this book. Through many drafts of the manuscript, she maintained a high degree of professionalism in making the changes. We sincerely appreciate her efforts.

We would like to thank Brother B. Austin Barry, Howard Epstein, Judy Hakola, P. Aarne Vesilind, and David L. Westerling for their comments and assistance on the preparation of the final manuscript.

CHAPTER 1 / **INTRODUCTION**

1.1 THE IMPORTANCE OF EFFECTIVE COMMUNICATION

In a recent discussion of the importance of technical writing skills, a partner of an engineering consulting firm explained that new employees with deficient writing skills are enrolled in a short course on technical writing. An employee who does not develop adequate writing skills after six months is discharged. The engineer believes good writing skills are that important in professional practice. Technical ability alone is not sufficient justification for keeping someone employed. Poor writing can lead to critical errors and present a poor image of the company to clients.

While this is just the view of one person, the results of numerous polls support this viewpoint. A study by Kimel and Monsees (1979) indicated that engineering practitioners believed communication skills are the most important expertise in engineering practice but that those skills are most often deficient among engineering graduates. As Table 1 shows, approximately 60 percent of the civil engineering respondents rated communication skills as being very important, and 65 percent indicated that recent civil engineering graduates lacked communication skills. Kimel and Monsees presented similar results for electrical and mechanical engineers.

Davis (1977a, 1977b) conducted a study to determine how much time prominent engineers spent either writing or working with materials that others have written, how important it is in their positions, and how important the ability to write effectively might be to someone who is being considered for advancement. The results of a questionnaire, completed by 245 prominent engineers, indicated that they spent an average of 24 percent of their time writing and 31 percent of their time working directly with material that others have written. Further, the respondents believed that their writing is important, often critical, to their positions and that young engineers often don't write well.

An ASCE study (1990) agreed with the findings of Kimel and Monsees (1979). As indicated in Table 2, graduates received the poorest ratings on report writing and oral communication. The ASCE study also ranked the writing skills considered most important in engineering practice: clarity, organization, grammar, and sentence structure (Table 3).

1

TABLE 1.   Capabilities of Recent Civil Engineering Graduates (Kimel and Monsees, 1979)

Importance to civil engineer practice				Capability of recent c.e. graduates (1-5 years)		
Most Imp.	Impor-tant	Less Imp.		Super-ior	Ade-quate	Infer-ior
			AREA OF COMPETENCE			
137	86	9	Writing and speaking	7	69	142
106	127	9	Structural analysis and design	25	160	28
72	156	13	Soil mechanics and foundations	9	157	46
62	125	47	Water and waste-water treatment	13	144	27
55	131	38	Fluid mechanics, hydraulics, hydrology	10	136	42
53	153	32	Computer and numerical methods	42	139	30
50	132	49	Economics, finance	4	96	106
47	147	54	Construction methods and equipment	9	107	95
46	105	70	Law, labor, management	4	81	120
20	148	67	Surveying and measurement	4	139	65
19	110	92	Transportation, highways, traffic	5	161	30
14	124	51	Materials	6	153	35
13	85	132	Social sciences and humanities	12	137	53

1.2   OBSTACLES TO EFFECTIVE COMMUNICATION

In professional situations, written and oral communication must be clear and concise so that the reading or listening audience receives the proper message. As a comparison, imagine that you tune to radio station QNOT. The reception is fuzzy. You listen intensely for a short while, but you keep missing words; you can't quite understand what the sponsor is trying to sell you; finally, you lose patience and switch to QYES. The reception for QYES is clear. You understand what the announcer and sponsors are saying and can enjoy the program being broadcast. So you continue to listen to QYES. Since you no longer tune your radio to QNOT, the sponsors on QNOT will not have the opportunity to present their products to you.

TABLE 2. Evaluation of Education Preparedness of Recent Bachelor Degree Graduates by Discipline (ASCE, 1990)

Discipline	Mean response[a]		Percent inadequate	
	Educator	Practitioner	Educator	Practitioner
(A) ENGINEERING AND SCIENCE				
Structures	1.13	1.43	1.6	7.2
Soil mechanics	1.19	1.62	9.8	3.8
Physics	1.22	1.29	2.4	2.6
Mathematics	1.26	1.20	1.0	4.2
Surveying	1.32	1.75	19.1	6.2
Chemistry	1.40	1.40	5.0	5.2
Computer Science	1.43	1.43	5.7	5.7
Thermodynamics	1.47	1.57	14.3	9.7
Electrical circuits	1.52	1.64	17.5	10.9
Engineering drawing	1.56	1.85	26.7	12.5
Statistics	1.88	1.76	14.4	23.2
Biology	2.07	1.62	27.0	40.1
Mean	1.45	1.55		
(B) WATER				
Fluid mechanics	1.14	1.64	5.1	2.6
Hydrology	1.32	1.59	13.8	4.7
Environmental	1.33	1.64	10.8	4.2
Water resources	1.40	1.59	6.9	4.8
Mean	1.30	1.62		
(C) GENERAL				
Economics/finance	1.65	2.14	33.5	9.5
Social	1.93	2.00	26.4	25.0
Report writing	2.05	2.48	42.4	27.5
Oral communication	2.11	2.44	43.8	30.0
Contracts/legal	2.21	2.27	47.2	38.9
Mean	1.99	2.27		

[a]Scale: 1 = Sufficient, 2 = Marginal, and 3 = Inadequate.

The desire for clear broadcasting is similar to communication in professional situations. You want your audience to receive your entire message. If your oral presentation is not clear or if your communication is not concise, your audience may lose interest and switch you off. Just as it is important for sponsors to get their message across to the listening audience, it is important for you, the professional, to get your views across to the client.

TABLE 3. Emphasis Needed in Technical Writing Skills (ASCE, 1990)

Skill	Average response[a]	
	Educators	Practitioners
Clarity of thought	1.1	1.1
Report organization	1.6	1.7
Grammar and syntax	1.9	2.0
Sentence structure	1.9	2.0
Vocabulary and spelling	2.0	2.1
Rewriting of drafts	2.2	2.5
Use of visual aids	2.3	2.5
Writing for nontechnical audiences	2.4	2.2
Letters and memos	2.6	2.4

[a]Based upon scale of 1 (essential) to 5 (unimportant).

Techniques for effective oral and written communication are essential tools in any business. When you send your employer or an employee a memorandum or a report, you want that person to understand exactly what you meant to say, not to interpret the message in some other way. When you present a summary of your work either to your superiors or to clients, they will be interested and receptive to your message only if you speak in a clear, concise manner. A clear transmission of messages maximizes efficiency and enables those receiving the message to respond properly.

What prevents someone from becoming an effective communicator? What are the obstacles to effective communication? If we asked "What are the obstacles to a football player being successful on the field?" we would probably identify such factors as: attitude, an unwillingness to practice, a lack of skills and knowledge, a lack of concern about the opponent so that an effective game plan isn't developed, a lack of concentration so that the plays are not executed properly, and an unwillingness to evaluate past performances. These obstacles to winning in sports also prevent the professional from being an effective communicator.

The first obstacle to effective communication is attitude. When people have problems with their communication responsibilities, they often rationalize that it is the technical solution of their work that should be judged, not their communication ability; they are not English majors. This is a problem of attitude. Both the student and the professional should realize that the development of their communication skills is just as much a professional responsibility as the mastery of technical skills. To overcome the obstacle of a poor attitude, it is necessary to acknowledge that the ability to communicate is a primary determinant of movement up the corporate ladder.

Coaches often attribute poor performance of a player during the foot-ball season to the player's unwillingness to practice during pre-season drills. Just as athletes avoid practices, students avoid courses where their communication skills will be enhanced; they avoid classes where they will be required to write papers and make oral presentations. The unwilling-ness to practice, whether in the football pre-season or in classes that require the use of communication skills, is a serious obstacle to effective communication. It may be partially responsible for the findings of Kimel and Monsees that recent engineering graduates generally lack good com-munication skills. Practice is the solution to this obstacle.

The lack of adequate skills by players at certain positions is an obstacle faced by football coaches when developing plays. Similarly, a speaker or writer must give proper consideration to the skills of the listening or reading audience. Every communication must be tailored to the abilities and characteristics of the intended audience. For example, if you don't know your listening or reading audience, you may be giving them a message that is too technical, too advanced, or too detailed. A message clear to an experienced group may be lost on a less experienced one. Recognizing the importance of assessing the intended audience is one of the most im-portant obstacles to overcome when formulating a communication.

Football coaches organize game plans to prepare for games. The in-ability or unwillingness of a professional to organize communication can be an obstacle to effective communication with an outcome similar to improper organization in football. For written or an oral report, this prep-aration takes the form of a series of outlines. While outlining requires time, skipping the progressive outlining step will actually decrease your efficiency rather than improve it. The first outline—initial attempt at organization—should not be viewed as a hurdle to jump but as a tool that gives direction to the formulation of the report or speech.

Another common mistake is to fail to view your report or speech as a whole. Taking the "big-picture" helps you make sure that the words in a sentence, the sentences in a paragraph, the paragraphs in a section, and the sections in the entire report are properly coordinated. To be understood, thoughts and ideas must be properly connected by some transition. Without these transitions, the audience will find it difficult to follow your ideas.

Getting started is the most common obstacle to effective communi-cation. Communication responsibilities are easy to put off, especially the major ones. The completed report or speech, seems far removed from the initial attempts at outlining and this provides an excuse for not getting started. Time spent on the intermediate steps (the outlines and rough drafts) is often viewed as wasted time. This view encourages procrasti-nation since any delay in getting started will reduce the time spent on the intermediate steps. Overcoming procrastination and starting prepa-ration early actually results in more effective communication.

Poorly prepared reports or speeches often present a confusing message, resulting in negative feedback for the author. Negative feedback, in turn, gives the author a negative attitude toward future speeches or reports. Attempts at communication are cumulative, not independent. Positive experiences lead to other positive experiences; negative experiences lead to other negative experiences. It is important to overcome obstacles to communication to ensure that the negative experiences do not accumulate. Now is the time to begin on the path toward positive communication experiences.

The purpose of this book is to help both college students and beginning professionals overcome these obstacles and learn to communicate effectively through writing and speaking. The book provides guidance in writing reports, memoranda, resumes, and business correspondence. You will learn to organize your thoughts and words into an appropriate written structure and overcome common roadblocks to effective communication.

This book will also be a guide to improving oral communication to both an audience and an individual. Clear oral communication is essential in person-to-person interactions, such as a job interview. You will be guided in both organizing and verbalizing thoughts and preparing effective visual aids to present technical information to a group. Body language as part of speaking with or to others is also discussed. Overcoming the obstacles associated with these aspects of communication in professional life will help you advance more rapidly in professional life.

What is the best way to learn to communicate more effectively? Do it, and do it a lot. The more one practices, the easier it becomes. While in college, seek out courses and other opportunities that require and enhance communication skills. Courses that involve written reports and oral presentations should be sought, not feared. Participate as an officer in student organizations because it is an opportunity to develop leadership and communication skills. As entering professionals, students should become familiar with the structure of written communications in the company and the methods used to meet oral communication responsibilities. These can be some of the best opportunities for developing effective oral and written communication skills.

In addition to practicing, take every opportunity to read and listen to others speak. Read as many reports or other written materials as time permits. Think about what makes the report readable. Is it well organized? Are the headings adequate, descriptive? Is the sequence of paragraphs logical? If the report is poorly written, determine why. How would you revise the report so that it would communicate the message better?

Listening to others speak can similarly help improve your oral communications. Go to conferences in your field or meetings or seminars at your school or workplace. Pay close attention to the speaking techniques of the various presenters. Jot down some notes on these talks; the notes will be helpful when you prepare a presentation. If the speaker is bad, note what he or she has done to confuse or bore you. Does the speaker talk in a monotonic voice? Does he or she speak without any body movement? Is the speech disorganized? If the speech is good, note how the speaker has caught your attention. Are the visual aids interesting? Does the speaker appear to be confident? It is just as important to learn about oral communication as it is to learn the technical knowledge being presented.

If you practice communication and take notice of others as they communicate, both oral and written communication will become easier and more natural to you. Practice is the easiest and best way to overcome obstacles to effective communication. Then when it counts, you will be able to communicate clearly and effectively.

1.3 CRITERIA FOR JUDGING EFFECTIVE COMMUNICATION

What is meant by the term effective communication? Effective means *serving the purpose* or *to produce the desired result*. In communicating technical information, the desired result may be to educate the reader or the listener. For example, a researcher orally presenting research results will be judged effective only if, at the completion of the presentation, the audience understands the research methods used, the major findings, and the implications of the results. The purpose of a communication may also be to persuade, such as a resume sent to potential employers. The cover letter and resume are an effective communication only if they persuade the personnel director to grant a job interview.

Entertainment is another objective of communication. Speakers often begin a speech with either a brief story that makes a point or a series of 35-mm slides that catches the audience's interest. Either of these is entertaining and can entice the audience to pay attention. They serve an important purpose of communication.

Skill development or instruction may be the objective of a communication, such as when a computer programmer compiles a user's manual for a software package. The communication will be judged effective only if users can apply the software and make very few errors. Whether the purpose is education, persuasion, entertainment, or instruction (i.e., training), the effort put into the communication is worthwhile only when the communication is judged to be effective.

Unfortunately, effectiveness is not a yes or no quality. A series of 35-mm slides shown at the beginning of a presentation may be entertaining to a few, boring to others, and mildly interesting to the rest of the audience. Thus, the effectiveness of the slide show was quite variable. This demonstrated that a set of specific criteria are needed to judge the effectiveness of a communication.

The news media, both print and television, is criticized frequently for biased inaccurate reporting. Some reporters are labeled biased toward the liberal viewpoint, while others are seen as promoting only conservative views. The media is also judged as being inaccurate; television concentrates on the entertainment aspects of a story, which leads some to charge that the stories are an inaccurate representation of reality. Television news is given a certain time slot into which all of the important news and entertainment, as well as the commercials, must be shoehorned. This leads to short reports artificially packaged into "newsbites" that are then criticized as being too brief.

The media argues that short reports are necessary and reflect an efficient transmission of the important aspects of news stories. That television news is efficient can be seen in its high ratings. Thus, primary criteria for judging the effectiveness of news communications are unbiasedness, accuracy, and efficiency. These same criteria apply to technical communication.

Unbiasedness usually means fair or evenhanded. A biased communication presents a one-sided view. For example, if a researcher made five experiments and presented only those that supported his or her viewpoint, the communication would be biased; such research would also be considered unethical. If an author is comparing alternative design methods and concentrates the discussion on the positive aspects of one alternative and the negative aspects of another alternative, then the writing is biased. A reader who detects the bias will be critical of the communication and will argue that it is not effective for the purpose of education. Thus, bias leads to ineffective communication.

Accuracy is another standard used to judge effectiveness. A communication that describes results that are as close to reality as possible is accurate. Accuracy has several aspects. First, the material should be proofread so that the number 25.3 appears as 25.3 and not 52.3. Second, numbers should be reported such that the digits shown in a final computation are significant; the topic of significant digits is discussed in Chapter 4. Third, accuracy in communication can be assessed in terms of grammatical correctness. For example, the following sentence can lead to an inaccurate interpretation: While they were feeding on grass, the park rangers rounded up the bison. If poor grammar leads to an inaccurate assessment, then the communication is not effective. To be accurate, communication must be structurally and grammatically correct.

Efficiency is another important criterion of effective communication. Time is important to professionals; they do not have it to waste. Therefore, material will be judged on the basis of the effort required to obtain a unit of information. As long as conciseness is not achieved at the expense of accuracy and completeness, effectiveness is directly related to efficiency.

In summary, effective communication skills are just as important as technical ability to the professional. It is not sufficient to complete the technical requirements of a project. The engineer must communicate the work to the client or other professionals. The communication must be accurate and unbiased, as well as completed in an efficient manner.

1.4 THE SCIENTIFIC METHOD IN EFFECTIVE COMMUNICATION

We have explained that effective communication is critical to one's success in professional life. We recognize that there are obstacles to effective communication including attitude and a hesitancy to approach communication responsibilities using a consistent, organized strategy. The development of a personal problem-solving philosophy about communication responsibilities is an important step in becoming an effective communicator.

The process approach to writing became popular about 10 years ago. In this approach, favored by many language-arts teachers, there are four phases: planning, writing, revision, and publication. The planning phase involves developing the idea and gathering pertinent background information. In the writing phase, the "discovery" draft is written; this is the initial effort aimed at organizing the major points into a systematic framework. The third phase involves editing, both self-editing and editorial reviews by others. In an engineering office, this will usually involve editing by the project manager as well as others who have worked on the project. Finally, the document is placed in final form, the publication phase. Everyone goes through these general steps during the writing process or in developing an oral report.

Engineers are familiar with the scientific method of investigation. It has been credited with many scientific advances, although in many cases the element of chance was very evident in the success of the discovery. But it is generally believed that following the steps of the scientific method increases the chance of success.

While there are many formulations of the scientific method, we will assume that it consists of the following four steps: observation, hypothesis formation, experimentation, and validation. In the first step, the engineer recognizes a problem or observes some regularity in a physical phenomenon that needs explanation. Based on this observation, the engineer formulates a hypothesis that reflects a cause for the regularity or effect. In the experimentation step, a controlled plan is developed and the experimental plan is conducted for testing the correctness of the hypothesis. Finally, the concept behind the hypothesis is verified by comparison to other observations, and it is used as a generalization to explain similar phenomena.

Can the systematic nature of the scientific method be used in a developing of a philosophy for effective communication? The observation step involves identifying the central theme to be communicated, including its extent, limitations, and the intended audience. Failure to place the appropriate importance on this phase will reduce the effectiveness of the communication. This phase would correspond to the planning phase of the process approach to writing.

The development of an outline corresponds to the hypothesis formulation stage of the scientific method. The outline is a statement of intent and sets down the scope of the work. If the outline, like a scientific hypothesis, is correctly formulated, then refinements in the subsequent steps of the process should yield an effective communication product.

The experimentation phase of the scientific method corresponds to the process of expanding the basic outline into more detailed outlines and a final manuscript. One would experiment with different examples, figures, and data to develop the general thesis of the communication. Editing to improve criteria such as accuracy and unbiasedness is part of this phase of the process.

When the final product has been developed, it will finally be tested on the intended audience and its effectiveness evaluated. If it is judged to be effective, then the report or presentation will be "validated."

Whether one approaches a communication responsibility with the process approach of the language-arts community or the scientific approach of the engineering profession, it is important for the individual engineer to develop an organized philosophy for completing communication responsibilities.

1.5 WRITING FOR A SPECIFIC AUDIENCE

If a restaurant consultant is asked what are the three critical factors for success, she will say: location, location, location. Similarly, if a writer is asked what the three critical factors in communication are she will say: audience, audience, audience. Once a document is designed and written with a specific individual or special group clearly in mind, effective communication will follow.

There are several questions that should be asked in order to address the audience of a document. Three are:

1. Who will read the document?
2. What technical background does the reader have?
3. Why does the reader want the document?

It is also important to be aware of any deadlines; late information, no matter how well written, is useless to the reader.

1.6 REFERENCES

ASCE Task Committee on Water Resources Education and Training (1990). "Perspectives on Water Resources Education and Training," *J. Water Resour. Plng. and Mgmt.*, ASCE, 116(1), 99–133.

Davis, R.M. (1977a). "Technical Writing: Who Needs It?" *Engineering Education*, ASEE, 67(2), 209–211.

Davis, R.M. 1(977b). "Technical Writing in Industry and Government." *J. Technical Writing and Communication*, Vol. 7(3), Baywood Publishing Co., Farmingdale, N.Y., 235–242.

Kimel, W.R. and J. Monsees. (1979). "Engineering Graduates: How Good Are They?" *Engineering Education*, ASEE, 70(2), 210–212.

CHAPTER 2 / **PROCRASTINATION**

DO

- admit to being a procrastinator—it is the first step to overcoming the problem
- assess your tendency to rationalize—identify your most frequently used excuses
- make lists and prioritize
- simplify difficult tasks and attack the individual subtasks
- set a time schedule

DON'T

- keep your tendency to procrastinate a secret—tell your friends and seek help
- delay tasks until tomorrow—DO IT NOW!
- ignore your successes when you did not procrastinate.

2.1 INTRODUCTION

Why is the topic of procrastination part of a book on communication? After all, the word *procrastination* means to postpone or delay needlessly while the word *communication* means to transmit or exchange ideas. Procrastination implies a hesitancy to act such as in the transmission of thoughts to paper or to someone. This is the very reason for addressing the subject of procrastination.

Procrastination reduces the efficiency and effectiveness of communication. On a written communication project, it prevents an individual from getting started, thus delaying the process. Procrastination also reduces the free flow of ideas that leads to creative, effective writing. In preparing an oral presentation, the individual who procrastinates will have to compress the time devoted to the first three stages of oral communication: formulating, developing, and rehearsing the presentation. Inevitably, this loss of time will result in a less polished presentation.

Delay is a hurdle that some people must jump over before they can become effective communicators. It can be detrimental in two ways. First, it can prevent one from successfully completing a current task. This leads to the second effect: Failure reduces confidence and self-esteem, thus making future communication tasks all the more difficult. Removing the procrastination hurdle will significantly improve one's attitude towards communication tasks.

Those with a history of procrastinating over communication tasks will not read a book on communication enthusiastically. Their negative attitude can actually decrease comprehension of the book. Thus, procrastination is discussed in this second chapter so that procrastinators can recognize their problem as solvable. Studying and practicing the methods in this chapter can help overcome the problem, thus enhancing the reader's understanding of the remaining chapters of this book.

2.2 STRATEGIES FOR OVERCOMING PROCRASTINATION: A BAKER'S DOZEN

Just as the M.B.A student learns methods of goal setting, planning, organizing, and problem solving, the procrastinator needs to know that there are strategies for overcoming procrastination. The same method cannot solve all bouts with procrastination. Different methods are appropriate for different situations, so being aware of alternative methods can increase one's success rate. He or she can then select the most effective method for the specific problem encountered. In the following paragraphs, 13 methods are briefly described. The judicious use of these methods can go a long way toward overcoming procrastination, thereby improving communication ability.

2.2.1 The Wish-List Method

At Christmas and Channukah, many children (and many adults) compile a wish list of anticipated presents. The list often includes some things that are needed and things that would bring pleasure. The list maker often ranks the items in the order of preference, so that those who have a copy of the list can give the gift(s) most desired.

The procrastinator can use the same concept to get started. The procrastinator should make a list of all the tasks that must be completed within the short-term; this includes both dreaded tasks and pleasurable activities. Next to each task in the list, make two columns. In the first, rank the tasks in order of importance, with 1 for the most urgent task and n for the least urgent. In the second column, rank the tasks in order of the degree of pleasure that their completion will provide, with a rank of 1 for the most pleasurable and a rank of n for the least pleasurable. Now sum the two ranks. Assuming that there are no immediate deadlines on any of these tasks, the tasks can be selected based on the summed ranks. For example, some may prefer to start with the lowest sum, others with the highest sum. Completing tasks on the list will reduce the tendency to procrastinate since accomplishment usually breeds satisfaction.

Fig. 1 provides an example of a student's list for weekend tasks. The eight items include some that are necessary and some that aren't, some that are pleasurable and some that aren't. After rating each task on the two scales and summing the ranks, it is evident that the soccer game at 1 o'clock is first priority. But until it is necessary to leave for the game, the student should spend the time studying for the calculus quiz. Following the soccer game, the student should read the history assignment, finish the laboratory report for chemistry, and do the laundry. Outlining the ENGL 101 paper, paying the phone bill, and balancing the checkbook have low priority. Attacking the tasks according to the priority identified by the total score helps the student focus on the important tasks while maximizing pleasure to the extent possible.

2.2.2 The Alfred Hitchcock Method

In the early years of TV, a show titled "Alfred Hitchcock Presents" started by showing an outline of Alfred Hitchcock. Viewers didn't require the spoken introduction to know what was coming; the outline of Hitchcock triggered the antipication of what was coming—a suspenseful show.

When preparing a written report or formulating an oral presentation, an outline will set the writer's mind in the proper perspective, just the way that the outline of Alfred Hitchcock prepared the viewer for suspense. It is best to start with a very brief outline and develop more detailed outlines progressively. The use of outline form forces the writer to focus on the important topics. It also helps to organize the ideas into a systematic form. The progressive outline approach is discussed in more detail in Chapter 6.

Task/Activity	Importance	Pleasure	Total
do laundry	2	7	9
finish chem lab report	6	2	8
soccer game @ 1 p.m.	3	1	4
study for calculus quiz	1	4	5
outline paper for ENGL 101	7	5	12
read history assignment	4	3	7
pay phone bill	5	7	12
balance checkbook	8	7	15

FIG. 1.—*Wish-List Method*

2.2.3 The Nagging Spouse Method

The classic situation on a family comedy show involves the nagging spouse. For example, the wife may nag the husband to take out the garbage or clean out the garage. The husband may nag the wife to fix dinner. The parent may nag the child to clean his or her room. It is assumed that the nagging produces results. If it works for the nagging spouse or parent, it can work for you. When you find yourself procrastinating, nag yourself about it. Many people have learned that if they go ahead and do something they avoid being nagged. So get into the habit of nagging yourself when you are procrastinating, then you will stop putting the task off if only to avoid your own nagging.

One technique for self-nagging involves posting notes to yourself in conspicuous places: phone receiver, computer, bathroom mirror, car, or entertainment center. A nagging note on the VCR or television has the added advantage of suggesting a reward once the task is completed. It is most important not to put a task out of mind. Nagging keeps it in focus.

2.2.4 The Titanic Method

Unsuccessful attempts to overcome procrastination can serve as either positive or negative motivators. If you take the attitude that these unsuccessful attempts prove an inferior ability, then the procrastination can sink your confidence and self-esteem. Instead of allowing past experiences to be the Titanic of your self-confidence, you should recall bad experiences of the past as motivators to avoid the procrastination that was an element of these occurrences. The memories can direct you toward positive action rather than negative inaction.

2.2.5 The Divide-and-Conquer Method

When tasks seem overwhelming, everyone has a tendency to avoid getting started. However, you can overcome the tendency by dividing the major task into a series of less "overwhelming" tasks and then set about completing the overall task by conquering the series of smaller tasks. For example, if you are having a party, cleaning the entire house might seem like an impossible chore. But if the house cleaning is viewed as a series of room cleanings, you can conquer the chore by attacking the individual rooms; cleaning a series of rooms seems far less overwhelming than cleaning an entire house. Similarly, it would be easy to put off the task of writing a 250-page thesis. But viewed as writing five 50-page chapters, it is less stressful. If each chapter is further broken down into sections, then the problem is more easily solved.

2.2.6 The Solitary-Confinement Method

The environment in which one must solve a problem can be a critical inhibitor of success: a procrastination-inducer. An environment with a lot of distractions can impede progress. A noisy dorm room is not conducive to writing a term paper. Committing yourself to solitary confinement in a little-traveled part of the library can help overcome a distracting environment.

2.2.7 The Town-Crier Method

Procrastination is easier if no one knows of your commitment to a task. Telling friends and relatives of your commitment to complete a task can serve as a pressure to get started. But don't tell just one person. Instead, like the town crier, tell everyone. Once everyone knows, then the fear of failure may serve as a motivator, not an inhibitor. Also, your considerate friends, knowing that you have a serious task at hand will hesitate to suggest distractions.

2.2.8 The Roll-the-Presses Method

The editor of a newspaper cannot procrastinate. When it's time for the presses to roll, the copy must be ready. Deadlines serve as motivators to overcome procrastination. Should deadlines on your communication tasks be flexible? Probably not! If the deadline is flexible, then it is easy to delay getting started because there will be a tendency to put the task off until the last possible moment. Many of us would put off our income tax if the IRS were flexible. But the April 15th deadline and the penalties are motivators to file our returns on time. So be like the editor and the IRS: set deadlines and practice meeting those deadlines. Establish self-imposed penalties for not meeting a deadline.

2.2.9 The Some-Enchanted-Evening Method

Just as recalling negative consequences of past bouts with procrastination can serve as motivation for overcoming it, thinking of past successes can also motivate by engendering good feelings about accomplishment. But you should not just recall past successes when confronted by a task. Past successes at avoiding procrastination should be recalled at other times to serve as positive reinforcement. So it is important to recall past successes when you are relaxing and enjoying a setting sun. Then the positive attitude will be more deeply embedded within the psyche.

2.2.10 The Downhill-Racer Method

When struggling with a task, procrastination feeds on itself. A confidence builder is needed. One avenue to success is to follow the downhill racer: get a fast start by attacking an easy part of the problem first. Save the hill climbing until some success has been achieved. The momentum from the downhill part of the race can help propel you through the uphill struggles. So select an easy part of the overall task and complete it before attacking the more difficult parts.

2.2.11   The Skinner Method

We are all motivated by rewards. In Psychology 101, the experiments of B.F. Skinner exemplify his hypothesis that behavior is determined by its consequences. Just as Skinner conditioned mice to push the levers when they wanted a food pellet, the procrastinator should reward himself or herself when a task is completed in a timely manner. In the early stages of tackling procrastination, Skinner's successive approximations approach can be used. Specifically, major changes in behavior are not to be expected, so any positive change in behavior, however small, should be rewarded. However, the change must be truly measurable, not the result of a rationalization that a partial success was achieved. Significant progress must be made before you give yourself a pizza, a movie, or whatever.

2.2.12   The 3 × 5-Card Method

Procrastination can result from the inability to focus on the real problem. Minor tasks should not take precedence over the actual issue. Instead, the main goal must be the focal point of your energies. A 3 × 5 card can be a focus facilitator. Writing the central problem or goal concisely will make the solution more apparent. A clear statement of your goal or thesis, propped in front of your computer, can help avoid procrastination.

2.2.13   The Procrastinators Anonymous Method

For some, other people can serve as motivators, especially when these other people share the same problem. Alcoholics Anonymous and dieting groups are successful because people with the same problem share a common goal. Procrastinators share a common problem, and working on communication assignments at the same time and place with other procrastinators can reinforce individual efforts. One way to do this is to help each other select, and stick to, appropriate strategies.

2.3   RATIONALIZATION AND PROCRASTINATION

Rationalization is an attempt to falsify the results of one's action to avoid regretting the action. It serves to justify actions both to avoid guilt and to avoid changing one's value system. A drunk avoids changing behavior by arguing that alcohol is not a problem. Compulsive eaters convince themselves that a *small* piece of chocolate mousse cake cannot have many calories.

Procrastinators rationalize about their problem in much the same way. They argue that the assignment is not that important to the final grade; deadlines are made to be broken; everyone else will be late; and tomorrow I'll do a better job. These excuses enable the procrastinator to avoid regretting not starting the task.

To overcome the cycle of rationalizing, putting off the task and then failing, you must recognize the rationalization and its consequences. Once you seriously admit to rationalization, you can revise negative behaviors and eliminate procrastination.

2.4 EXERCISES

1. Identify several instances in the past when procastinating has led to undesirable results. Look for commonalities among the examples and select one of the 13 methods discussed in this chapter to help overcome the problem. Discuss how the method would have helped you to get started in at least one of the situations you cited.

2. Compose a wish list of the tasks involved in an upcoming assignment. Now compose a wish list of the tasks associated with an upcoming social activity. Which list is longer? What does this suggest to you?

3. Create a list of tasks and activities to be completed in the near future. Rank the items in your list according to both importance and pleasure. Add the ranks. Does the order of the total rank reflect your priorities? Complete the tasks according to some decision criterion. When all tasks have been completed, evaluate whether or not the listing and ranking improved your task-solving efficiency. Write out your evaluation.

4. List three future tasks on which you have procrastinated (or intend to procrastinate). For each task, list a self-imposed reward. Write the reward in bold, large letters. Display it in a prominent place until all three tasks are completed and you have rewarded yourself.

5. What rewards could induce you to not procrastinate? Create a list and use it whenever you find yourself procrastinating over a communication assignment.

6. As part of your physics course, you must write a laboratory report and make a 5-minute presentation on the laboratory dealing with electromotive force and circuits. Using the divide-and-conquer method, discuss how procrastination can be overcome in getting started on the assignment.

7. Your company is asked to bid on part of a land development project. You are assigned the task of preparing the proposal. Use the Alfred Hitchcock method to show how a tendency to procrastinate could be overcome.

8. Identify two or three people who you would not want to know about your tendency to procrastinate. Make a resolution to tell these individuals about your communication assignments so that you will be less likely to procrastinate.

9. Identify two or three specific locations that you believe would

inhibit your tendency to procrastinate. Explain why you believe these locations are conducive to completing projects.

10. Imagine that you are a graduate student who has recently completed six months of collecting laboratory data and analyzing the results. How could you use the downhill-racer method to overcome procrastinating about getting started on the task of writing the thesis? Which of the other methods could also be used? Explain.

11. Write a rationalization of why you should *not* do two future, dreaded tasks. Summon up all the reasons you can think of for not completing the tasks. Read your rationalization aloud to yourself or to your classmates. Have you convinced yourself?

12. List excuses you use when rationalizing about communication assignments.

CHAPTER 3 / TECHNICAL WRITING STYLE

DO

- build and revise vocabulary constantly
- use active, not passive voice
- use parallel structure to balance words, phrases, clauses, and sentences
- know both denotative and connotative meanings of words

DON'T

- begin paragraphs with weak, narrow topic sentences
- write run-on, comma-splice, or fragmented sentences
- use vague, abstract words

3.1 INTRODUCTION

Writing style, though difficult to describe, is integral to every written communication. Style is influenced by many elements: convention, culture, circumstances, values, and the psychological make-up of the writer. Style immediately labels a writer. The average individual can usually identify many writers after reading only a few sentences. Most readers could tell the difference between a passage by Ernest Hemingway and one from a Harlequin romance. Similarly, few would confuse a sonnet by Shakespeare with a poem by e.e. cummings. How does one make these stylistic calls? Vocabulary, word choice, sentence structure, and punctuation are some of the elements, in addition to the less tangible components mentioned above, which help identify a poet or novelist.

Technical writing style is a product not of personal, artistic idiosyncrasies, but of convention and circumstances. The ideas, values, and background of the writer should not interfere with the communication of information to the reader. Nevertheless, all of the components of creative style—vocabulary, sentence structure, punctuation, and paragraph organization—must be understood in order to write clear, concise, efficient technical documents. Here are seven rules for effective writing:

1. Decide what you want to say.
2. Keep your audience in mind.
3. Make an outline. It will keep you on track and will reduce writing time.
4. Write sentences that are clear and effective.
5. Use simple, direct words.
6. Write as concisely as possible.
7. Check to make sure that grammar and punctuation are correct.

There are many excellent handbooks and even videotapes for writers that give detailed information on grammar and punctuation and provide numerous practice exercises. The purpose of this chapter is not to teach fundamental English usage but to explain how the three building blocks of technical writing can be mastered to produce an effective, appropriate writing style. The three building blocks are: words, sentences, and paragraphs. These will be discussed in the following sections.

3.2 WORDS

A broad vocabulary is essential to the efficient and concise use of words. Many students use too many words because they are unaware that there is one word into which all the rest can be subsumed. Word choice is the result of careful thought and revision. Pascal, the French philosopher, once dashed off a 10-page letter to a friend and concluded with these words: "Forgive the length of this letter; I did not have the time to make it shorter." In today's business world, many companies are demanding shorter, more concise writing. A past president of Proctor and Gamble instituted the "one-page memorandum." Longer memoranda were instantly returned to the writer with a demand to "boil it down."

The two keys to effective word choice are vocabulary building and revision. The best way to build your vocabulary is, of course, to read. Unfortunately, many people do not have the time or energy to read the type of books that would enhance their vocabulary. Another possibility is to use self-help vocabulary builder books that are widely available.

Make it a personal challenge to tackle a group of new words every week. Learn about their origins and linguistic roots and practice using them correctly in conversation. Understanding the origin of a word will help you incorporate it into your active vocabulary. Also, when reading, make it a practice to look up every word that you do not know.

Once you establish a good vocabulary, your next step is to get into the habit of revising your writing. Do not limit initial drafts by excessive concern with word choice. You can draw a line to indicate a space for the right word and later, supply, delete, or change words.

3.2.1 Denotation and Connotation

To write with clarity and precision, the student must recognize that every word communicates on two levels: denotative and connotative. Denotation means the dictionary definition of a word. Connotation means the emotional response associated with a word. Connotation, in turn, often depends on the age, culture, and history of the audience. For example, the word "party" means a gathering for a purpose. However, a group of six-year olds would have one connotation of the word, while a group of university students in a political science class would have another. Often, two words that come from the same root have very different connotations. For example:

> The fireman was praised for his heroism.
> The fireman was condemned for his heroics.

Heroism has a positive connotation, while heroics connotes the taking of a senseless risk.

Writers must be sure that all of the words in a sentence match up on a connotative level. For example:

> The timid little man sidled up to the policeman and *glared* at him.
> The timid little man sidled up to the policeman and *looked hopefully* at him.

In the first sentence, "glared" does not match "timid" and "sidled."

3.2.2 Concrete versus Abstract Words

Student writing is often marred by a reliance on vague, abstract words. Abstract words, being less specific, are used because they come more easily to mind. Abstract words describe qualities, ideas, or concepts. Examples of abstract words are: sweetness, love, education, poverty, honor. There are, of course, degrees of abstractness. The more specific writing is, the better it communicates. Compare the following pairs of concrete and abstract words:

Abstract	Concrete
The dessert had a sweet taste.	The cake was flavored with molasses.
Jane is reckless.	Jane ran a red light and two stop signs.
Mary enjoys reading.	Mary likes to read *Sports Illustrated*, *Vogue*, and Charles Dickens.
Plasma has a high flame temperature.	Plasma has a flame temperature of 20,000°F or 11,000°C.

The more writing is revised, the more specific and less abstract it becomes. Therefore, time should be set aside for several revisions.

3.3 SENTENCES

Sentence construction is a convention on which a culture "agrees to agree" in order to standardize and facilitate communication. Just as organized sport requires the players to agree on the regulations, so does organized communication demand all participants to agree on rather arbitrary rules. Sentence construction in English is not extremely complex, but there are five sentence errors that even well educated, well read individuals sometimes make. These are sentence fragments, run-on sentences, comma splices, errors in passive voice, and parallel structure.

3.3.1 Sentence Fragments

All sentences must consist of a subject and a verb. Sentences begin with a capital letter and end with a period, question mark, or exclamation point. Often, a group of words is punctuated so that it appears to be a sentence, but because it lacks a subject or a verb, it is considered a fragment rather than a sentence. Table 4 illustrates the difference between complete sentences and sentence fragments. What parts of speech are underlined in each complete sentence? How does the third fragment differ from the other two? What is the problem in the following example?

Because the conductor obeyed Ohm's law.

Although the group of words includes a subject (the conductor) and a verb (obeyed), the word *because* makes the sentence a dependent clause. The words *because, however, although, if, since, while,* and *until* should be avoided at the beginning of sentences unless they are followed by an independent clause; beginning a sentence with a dependent clause makes the words that follow dependent on an additional idea to make sense. Is this a sentence or a fragment?

Although Mary is currently an engineering technician, she hopes to become an engineeer.

TABLE 4. Sentence Fragments versus Complete Sentences

Sentence fragment	Complete sentence
Extensive computer work required.	Extensive computer work *is* required.
Requires larger reservoirs than peak discharge control.	*Erosion Control* requires larger reservoirs than peak discharge control.
Also plotted using a rectilinear axis system.	*The values of v and i were* also plotted using a rectilinear axis system.

This group of words is a sentence. The first part of the sentence is correctly subordinated to the second part.

3.3.2   Run-on Sentences

A run-on sentence is the opposite of a sentence fragment in that it consists of two independent, complete sentences that are not separated by a period. The following are examples of run-on sentences:

> Sediment is the only water quality parameter evaluated the case study uses a number of site-specific assumptions.
> All engineers are required to attend the employees' benefits meeting this will be held in the West Conference Room.

Run-on sentences are easily corrected by adding a period and correct capitalization.

3.3.3   Comma Splices

Comma splices are encountered more frequently than run-on sentences, perhaps because punctuation is used between independent thoughts; however, the punctuation chosen, the comma, is incorrect. The following are examples of comma-spliced sentences:

> The cafeteria will be closed for repairs for the next two weeks, it is hoped that this will not cause serious inconvenience.

> The equipment provides a flexible signal acquisition and data analysis system, the acquisition interface can be programmed in languages such as BASIC or FORTRAN for almost any digital analog input/output disk.

In both of these examples, the problem can be corrected either by replacing the comma with a semicolon or by forming two sentences, with the comma replaced by a period and the next word capitalized.

3.3.4   Active versus Passive Voice

The term *active voice* means that the subject of the sentence is doing the action. In the sentence, "The rat ate the cheese," the rat, who takes the action, is the subject of the sentence.

The term *passive voice* means that the subject of the sentence is *not* doing the action. The acting force is buried somewhere at the end of the sentence. In the sentence "The cheese was eaten by the rat," we see this passive voice pattern. The second sentence is more wordy and less forceful than the first sentence.

In English we commonly see the *who-does-what* pattern. This is the active voice pattern. It should be used much more frequently than the passive voice pattern. Active voice is more natural, more direct, and more concise.

3.3.5 Parallel Structure

Words, phrases, clauses, or sentences are parallel when their structure is balanced. Parallel form should be used to express ideas simply and logically. Balance a word with a word, a phrase with a phrase, a clause with a clause, a sentence with a sentence. The following are five examples of incorrect parallel structure, followed by a correctly written structure:

1. *Awkward*: People begin to feel as though they have no faces and are insignificant. (Two ideas are expressed. The problem is that one is a clause and the other is an adjective.)
   Parallel: People begin to feel *faceless* and *insignificant*.
2. *Awkward*: Condemned by the court and since he was also denied a new trial, he faced death. (A phrase is followed by a clause.)
   Parallel: *Condemned* by the court and *denied* a new trial, he faced death. (Two phrases.)
3. *Awkward*: Her voice was clear, deep, and with resonance. (Two adjectives and one prepositional phrase.)
   Parallel: Her voice was *clear*, *deep* and *resonant*. (Three adjectives.)
4. *Awkward*: To play baseball or going fishing is my preference. (Infinitive and gerund.)
   Parallel: *Playing baseball* or *going fishing* is my preference. (Two gerunds.)
   Parallel: *To play* baseball or *to go* fishing is my preference. (Two infinitives.)
5. *Awkward*: Engineering classes taught live on-campus are able to be simultaneously broadcast via microwave and can be delivered throughout the country via satellite, while videotaping is also possible.
   Parallel: Engineering classes taught live on-campus can be simultaneously videotaped, broadcast via microwave, and delivered by satellite throughout the country.

Parallel structure is frequently used in business writing. In a vertical list, make the first word in each item parallel. The following is a correctly structured vertical list:

The new analyst had three major responsibilities:

1 *Collecting* data from the regional offices.
2. *Verifying* data information sheets for completeness and accuracy.
3. *Totaling* the results from the regional offices.

All of the items in the list begin with a gerund (collecting, verifying, and totaling).

In contrast to the preceding list, the following group of job responsibilities is *not* in parallel form.

The Telecommunications Technician is responsibile for the following activities:

1. To monitor all equipment.
2. Supervising student employees.
3. Must attend campus-wide meetings on telecommunications.
4. Reports monthly to the Chief Engineer.

Note that the first responsibility is written in the infinitive ("to" form), the second is a gerund ("-ing" form), the third is a command, and so on.

In addition to using parallel structure in job descriptions, it is also extremely important to use it in resumes. A list of duties or job responsibilities looks more professional and is much easier to read when listed in parallel form. Chapter 10 provides examples of resumes which use parallel form.

3.4 PARAGRAPHS

Paragraphs are the third "building block" of effective technical writing. A paragraph is a group of sentences that develops a single thought or idea in a logical, coherent, and unified fashion and that is organized by means of a topic sentence. Following the topic sentence, which is usually the first sentence of the paragraph, several support sentences follow. These sentences explain and expand upon the central point made in the topic sentence. The paragraph is often ended with a concluding sentence, which emphasizes the point of the paragraph.

3.4.1 Paragraph Length

Unfortunately, we cannot say that every paragraph will have a set number of sentences. But this is actually an advantage in that it enhances writing style flexibility. The length of a paragraph is important because it greatly influences the clarity of the idea that underlies the paragraph. A paragraph that is too long can be difficult to follow and important concepts may be lost in the middle of the paragraph. It may not be possible to adequately develop an idea in a paragraph of two or three sentences, so short paragraphs may not be effective for getting the major point across. Additionally, a series of very short paragraphs can lead to a disorganized report.

Paragraphs will typically contain four to ten sentences, although paragraphs in brief business correspondence such as memos may contain fewer sentences. A paragraph describing a step-by-step procedure may have considerably more sentences, especially if long lists are included in the paragraph.

3.4.2 The Topic Sentence

The topic sentence has a special relationship to the other sentences in the paragraph. It is a contract that you as a writer establish with your reader. To fulfill the contract and satisfy the reader that your topic sentence is valid, you must ensure that every sentence in the paragraph supports the controlling idea expressed in the topic sentence. In other words, the topic sentence is like an umbrella under which all other sentences should fit. If any do not, your paragraph will not be unified and will not communicate effectively. On the other hand, if every sentence in the paragraph provides direct or indirect support for the controlling idea, your paragraph will be unified. Every expository paragraph must be unified to be effective.

Examine the following paragraph and decide which sentences do not contribute to the controlling idea:

> (1) Pilots are the primary cause of many aircraft accidents. (2) Ignoring their responsibilities, many pilots fail to perform their duties efficiently, and tragedy has too often been the needless result. (3) History records that many fatal accidents have occurred, for example, because pilots failed to listen to the advice of air traffic controllers who were in a position to warn them about impending disasters. (4) To become an air traffic controller, one must be extremely intelligent. (5) Sometimes pilots are overtired, and they neglect to take the precautions necessary to avoid accidents. (6) They may even be taking drugs that slow down their physical reactions. (7) As we all know, statistics have proved that the number of college students who abuse drugs is increasing at an alarming rate, and few of these students realize that if they continue to use drugs, they will never be able to enter a career in aviation. (8) Sometimes accidents occur through a malfunction in the plane's equipment. (9) A door may open during flight, or a tire may blow out as the plane takes off. (10) Pilots, of course, aren't responsible for accidents such as these. (11) Perhaps most startling is the fact that every year one or two air traffic accidents are caused by student pilots who attempt journeys beyond their capabilities and end up producing catastrophes which destroy life and property. (12) Because they don't employ student pilots, commercial airlines are the safest form of air transportation. (13) The next time you take a commercial flight, you should be sure to ask yourself the following questions: Does the pilot look happy and healthy? Does the plane seem sound and sturdy? What are the weather conditions outside?

Note that sentences 4, 7, 8, 9, 10, 12, and 13 do not provide direct support for the topic sentence.

3.4.3   Unity, Coherence, Logic

Unity, coherence, and logic are qualities essential to a good paragraph.

Unity means there is a clear, umbrella-like topic sentence, and all subsequent sentences relate to, expand on, or explain the topic sentence.

Coherence means a paragraph includes transitional words and phrases that help the reader move smoothly from sentence to sentence. Transitions include standard expressions such as: again, however, nevertheless, beyond, next, likewise, for instance, first, second, in summary, in conclusion. Repetitions of key words and phrases and the use of parallel structure also serve as transitions and provide paragraph coherence. Listen to the Reverend Martin Luther King's speech "I've Got A Dream"; it is an excellent example of coherence and effective transitions.

Logic means that the reader arranges the sentences in a paragraph in an order that makes sense and is consistent. Often a student will write a good topic sentence and include five or six supporting sentences with appropriate transitions, but the sentences are out of sequence. There are several ways to logically organize sentences. These include:

1. Physical or geographical sequence.
2. Chronological sequence.
3. Emphasis sequence.
4. Cause-and-effect sequence.
5. Compare-and-contrast sequence.

3.4.4   Sentence Sequencing In Paragraphs

Physical or geographical sequence means that an object or a scene is described in the order in which the reader would actually see it. The description could proceed from near to far, left to right, top to bottom, etc. The following is an example:

> Gaseous air pollutants enter the human system via the respiratory system, but the volume absorbed through the cell walls of the lungs is limited by the natural defenses of the respiratory system. Part of the pollution in the inhaled air is intercepted by the mucous lining of the nasal passage; this part is subsequently discharged. Hairlike cilia that line the nasal and bronchial passages remove an additional portion of the pollutants. Smaller particles pass into the lungs, where they can become lodged, some of which are dislodged and expelled by coughing. The health of the individual is a factor in determining the portion of the pollution intercepted by the natural defenses of the respiratory system.

In this paragraph, the sentences describe a physical sequence based on the flow of air, from the mucous lining to the nasal passages to the lungs.

The topic sentence in this example indicates the beginning and ending points of the physical system.

Chronological sequence means that steps or events are mentioned in the order in which they occurred or should occur. Manuals and cookbooks arrange paragraphs in chronological order. The following paragraph is an example:

> Changing the oil in your car is a simple procedure and can save you some time and money. The first step is to place a catch bucket, which holds about two gallons, underneath the outlet of the oil pan, which is the reservoir part of the crankcase. Then remove the plug and allow the waste oil to completely drain from the oil pan into the catch bucket. Then the oil filter should be removed, with the oil in the filter also poured into the catch bucket. A new oil filter is then installed, and the oil pan plug is replaced. The crankcase is then filled with new oil as specified in the owner's manual. As a final step, the waste oil should be disposed of properly.

In this example, the sentences describe the sequence that someone would follow to complete the task properly. Someone could reproduce the steps in order to complete the task. The intent of the paragraph would be confused if the sentence indicating that the plug be removed was placed before the sentence stating that the catch bucket should be placed under the reservoir.

Emphasis is somewhat difficult to grasp as it requires a judgment call on the part of the reader as to which sentence is most important, which less so, and so on. Usually, the most dramatic sentence should be placed as the last sentence of the paragraph.

A cause-and-effect sequence is a useful technique for organizing paragraphs. Usually, the cause is discussed in the topic sentence and the remainder of the paragraph is devoted to effects. The following paragraph is an example:

> Urbanization, which includes the transformation of forested land and open space to land with a significant fraction of impervious surface, causes significant changes to the flood runoff processes. The loss of forested lands decreases interception storage of rain. Grading of the developed lands decreases the volume of depression storage and reduces the permeability of the surface soil layer, which decreases the ability of water to enter subsurface storage. Overlaying large expanses of land with impervious materials, such as concrete and asphalt, decreases the volume of water entering subsurface storage and increases the volume of water that flows quickly to small streams and rivers. The changes in

the runoff processes caused by urbanization increase urban flooding and decrease groundwater recharge.

In this example, urbanization is the cause, while the decrease in interception, the decrease in depression storage, and the increase in direct flow are the effects.

An effect-and-cause sequence describes a significant effect in the topic sentence. The supporting sentences that follow discuss the causes. Paragraphs structured in this way are common in technical report writing since the result is of primary interest and decisions will be a function of the result, rather than the causes. The topic sentence of the following example, which is a revision of the example paragraph for the cause-and-effect sequence structure, centers on urban flooding, which is an effect:

> Urbanization has caused significant increases in urban flooding and decreases in groundwater recharge. The transformation of forested land and open space to land with a much higher fraction of impervious surfaces increases flooding because of (1) decreases in interception storage associated with the loss of forest cover; (2) decreases in infiltration rates associated with decreases in the permeability of the surface soil layer; and (3) decreases in depression storage associated with the land grading that takes place during construction. Additionally, the increase in impervious cover also decreases the infiltration of water, thus increasing the volume of water that passes directly to small streams.

In this example, increased urban flooding and decreases in recharge are the effects. The causes are the changes in interception, depression storage, and impervious cover.

Compare-and-contrast is another useful strategy. In general, both comparison (the similarity of two items) and contrast (their differences) should be used. There are two methods to do this. One is to discuss all of the similarities and then all of the differences. The other is to discuss one similarity followed by one difference, a second similarity followed by a second difference, and so on. The following is an example:

> When buying a new lawnmower, the suburban homeowner can choose between gas-powered and electric-powered mowers. Owners of gas-powered mowers must store gasoline, which is a safety problem not encountered by the owner of an electric-powered mower. An electric mower is lighter than a gas-powered mower, which makes mowing easier. But an electric mower requires the homeowner to drag along the extension cord, which limits the practical range of the mower to about 100 feet from an outlet. The gas mower is more flexible because the cord is not required, so it is easier to mow around obstacles such as trees

and bushes. The electric mower is easier to start, especially after long-term storage. The gas mower has the advantage that it can be used to cut wet grass. In summary, both types of mowers have significant advantages and disadvantages.

In this example, the sentences state contrasts between the two types of mowers. As an alternative, the paragraph could be structured to list all advantages of gas-powered mowers, then the disadvantages, and then sequentially the advantages and disadvantages of electric mowers.

3.4.5 One-Sentence Paragraphs

As a general rule, one-sentence paragraphs should be avoided. A paragraph presents a central idea in the first sentence. Successive sentences are essential to explaining and developing the first sentence. A one-sentence paragraph does not provide enough information to develop a complete thought or idea. In a few cases, a writer may wish to make a single point that doesn't need clarifying or supporting information. An advantage of a one-sentence paragraph is that it gets the reader's attention. Thus, it is appropriate for presenting the central objective of the written piece or a controversial conclusion. The following is a one-sentence paragraph used to emphasize the objective of a written report:

> The objective is to demonstrate clearly that rates of erosion in rivers immediately downstream of small reservoirs may actually increase after the construction of the reservoirs.

The following sentence illustrates the use of a one-sentence paragraph of a controversial conclusion:

> We conclude that the proposed policy would actually harm the environment rather than help it.

Except when using the one-sentence paragraph for emphasis, the more conventional multi-sentenced paragraph should be used.

3.4.6 Paragraph Checking

After completing a rough draft of a report, you should plan to go through several revisions. As will be pointed out in Chapter 6, it is common to have a specific purpose when making each revision. Checking for the proper structure of each paragraph should be the objective for one revision. The following aspects of paragraph structure should be checked:

1. Topic sentence clear.
2. Support sentences properly sequenced.
3. Concluding sentence appropriate.
4. Variation in sentence structure.
5. Sentences properly punctuated.

3.5   DEFINITION

Definition means a logical technique that reveals the meaning of a term or word. It enables the writer to set boundaries or limits that separate the term from any other term. Definition is an important communication skill, and one that requires practice and concentration. Particularly in technical fields, an inability to clearly define words, terms, processes, and materials can result in both expensive waste and duplication of effort. One of the biggest problems in student writing is a tendency to discuss words that have not been adequately defined. A good definition consists of a formal descriptive sentence plus as many supplemental explanatory techniques as required to make the subject clear.

3.5.1   Formal Sentence Definitions

A formal sentence definition consists of four components: species, verb, genus, and differentia. Two examples will be used to illustrate the components:

Species	Verb	Genus	Differentia
Acceleration	is	the rate of change	of velocity with respect to time.
A carburetor	is	a mixing chamber	used in gasoline engines to produce efficient vapor of fuel and air.

The species is the term to be defined. It can be used with or without the definite (the) or indefinite (a, an) articles. The genus is the general category or class into which the item fits, while the differentia comprise the characteristics that separate the species from all other in the genus. Because it is a sentence, a form of the verb "to be" must be used.

It is important to carefully limit the genus. The narrower the genus, the more exact the definition will be. Consider the following examples:

Species	Verb	Genus	Differentia
A rifle	is	a dangerous object	that shoots bullets.
A rifle	is	a firearm	with a long, narrow barrel.

Which gives a clearer definition of the term? Here is another example. What is wrong with this definition? How could you improve it?

Species	Verb	Genus	Differentia
A belt	is	an article of clothing	that holds up pants.

At the same time, the genus should not be stated too narrowly. In the following definition, what types of chairs would not be included, due to an excessively narrow genus?

Species	Verb	Genus	Differentia
A chair	is	a four-legged piece of furniture	on which people sit.

Be careful not to repeat any form of the species in the differentia. To do so is to construct a circular definition, which does not clarify the writer's meaning. Circular definitions would include the following:

Species	Verb	Genus	Differentia
Comatose	is	a state	of being in a coma.
Fertilization	is	the process	of fertilizing an object.

Avoid the words "when" or "where" after the verb to be. Using when or where prevents the clear statement of the genus as a noun phrase. Which definition is clearer?

Species	Verb	Genus	Differentia
A crypt	is where	people	are buried.
A crypt	is	a subterranean chamber	in which people are buried.

Not only is the first definition awkwardly written, but it could also describe cemeteries and mausoleums. Here is another example:

Species	Verb	Genus	Differentia
Osmosis	is when	fluid	passes through a membrane
Osmosis	is	a process	in which fluid passes through a membrane

Finally, consider the language skills and technical background of the audience when constructing a definition. Do not use needlessly complex language or overly long words. It is actually harder to write a short, simple definition than it is to compose a long, complicated one.

3.5.2 Supplementary Defining Strategies

Often it is impossible to define a complex term in only one sentence, no matter how carefully the sentence definition is constructed. Usually, supplementary defining strategies are added to the sentence definition. Always begin with the formal sentence definition. Supplementary strategies can be as brief as an additional sentence, or grow into an entire paragraph. The most frequently used supplementary defining strategies are:

1. Etymology/word origin.
2. Example.
3. Illustration.
4. Analogy.
5. Synonym.
6. Negative Definition.
7. Operational Definition.
8. Analysis.

Sometimes comparison/contrast and cause/effect are used as supplementary defining strategies.

Word origin is useful when an acronym (SCUBA, SONAR) or initials (N.A.T.O., D.O.D.) are used. What do SCUBA, SONAR, SNAFU, SWAT, and LASER stand for? Etymology is also effective when a word has an interesting historical background. Look up chauvinist and gerrymander. Do these words mean the same today as they did originally? A good dictionary always provides etymologies. Often students overlook this useful information. The most detailed etymologies are found in the *Oxford English Dictionary* (O.E.D.) located in most libraries.

Technical writers and engineers are usually comfortable about including illustrations (exploded diagrams, maps, graphs, photographs, etc.) to supplement their definitions. Even a simple line drawing can help clarify definitions.

Analogies are very useful when a technical term must be defined for a nontechnical audience. An analogy points out a similarity between a familiar and an unfamiliar object. For example, a chemical equation is analogous to a recipe. A heart is similar to a pump. A turbocharger is like a bellows. Analogy allows the writer to be creative and imaginative.

Synonyms are also useful to explain a technical term to a nontechnical audience. The use of a familiar word that means basically the same as a technical word is always appreciated by the audience. For example, spun glass might be more understandable than fiberglass, voltage is more recognizable than electromotive force, and speakers are more familiar than electroacoustic transducers.

Often, a negative definition can help clarify a formal sentence definition. This technique is useful when a popular assumption is incorrect. Negative definitions must always be followed by a positive statement, such as an operational definition. The following are examples of negative definitions: "Italian food is not all tomatoes and garlic"; "Affirmative Action is not a quota system"; "A construction technologist is not an architect."

An operational definition tells what an object *does*. Operational definitions are frequently placed directly after a negative definition, or immediately behind a formal sentence definition. An operational definition can be acted on; its composition includes a simple test by which the term can be measured. Examples of operational definitions are: "Yeast (species) makes dough rise"; "A successful president (species) is one who holds the unemployment rate under 6%."

Analysis is very useful in defining technical terms. Analysis divides a complex term into parts and discusses each part. An operational definition may be added to explain what each part does. A definition of a sewing machine might break down into needle, bobbin, threading apparatus, tension control, feed-dog, etc. A definition of dynamics might necessitate an analysis of both kinetics and kinematics.

Read the following definition of sonar. Label the parts of the formal sentence definition, and underline and describe each of the supplementary strategies. To what type of audience is the definition directed?

AN EXPANDED DEFINITION OF SONAR

Sonar is a sensing device that uses sound waves to determine how far away an object is. Ships use sonar to assist in navigation by sensing the depth of water beneath them. Sonar is also used in fishing, military, and industrial applications.

Sonar is the acronym for SOund Navigation and Ranging. Nature has a sonar system called echolocation. Dolphins use this system to navigate through water; bats use it to navigate through air. The British Navy developed sonar after World War I because of the success of Germany's submarine operations. Britain first used it in 1921. It was called ASDIC because the Anti Submarine Detection Investigation Committee developed the first device. The United States developed sonar technology shortly after, and first applied it in 1927. Sonar was a closely guarded secret until World War II. As a result, the Allies enjoyed better success against the German Navy because of the element of surprise. After World War II many other nations developed sonar devices.

Sonar systems are comprised of four major components. First, a transmitting device generates an electric pulse. Second, a transducer converts the electric pulse into a sound pulse and vice versa when the pulse returns. The third and fourth components are a hydrophone, which is a sensitive microphone capable of listening in water, and a chip that calculates the time travelled and converts it to a numerical value representing the distance.

A sonar system sends a sound pulse through the water that is reflected back to the system after it reaches an obstacle. The resulting echo is received and the time is converted into the distance between the object and the ship. A submarine equipped with sonar will use it to detect potential obstacles or other submarines in the area.

Because sound waves are invisible, it may help to visualize a tennis match. Imagine the tennis ball is a sound pulse and the rackets are a transducer and an obstacle. The player serving the ball (the transducer) sends it over the net towards the other player's racket (an obstacle). If the second player returns the volley (reflects the sound wave), the ball will be returned to the first player (the transducer), who then receives the ball (sound wave) and converts it to the distance travelled.

Other applications of sonar technology include:

Photography: used to focus the camera lens based on the distance between the subject of the picture and the camera.

Ultrasound: used for non-invasive medical examinations including monitoring fetal development.

Industrial: used for measuring the level of liquid in a tank, liquid flow, liquid interfaces, and counting parts on conveyor belts.

3.6 EXERCISES

1. Rewrite each of the 10 sentences correctly.
   a. I respect my parents, I resent their attempts to choose a career for me.
   b. Why did so many New Yorkers support England during the Revolution this is a hard question to answer.
   c. Many states have passed the Equal Rights Amendment, however, there are several remaining states in which opposition to the amendment is strong.
   d. Scopes was prosecuted for teaching Darwin's theories, he was defended by Clarence Darrow.
   e. The candy bar in my glove compartment had melted sticky chocolate sauce covered my license and registration.
   f. After accepting the award, she cried joyously. Because she had never received any compliments for her work before now.
   g. She spent her first week on the job as researcher. Selecting and compiling technical information from digests and journals.
   h. If you decide to leave the activity at any time. Press the Return Key and the system takes you back to the main menu.
   i. When the waitress came to take our order, she listed the specials of the day. They included roast beef, fried chicken, Virginia-baked ham, and broiled swordfish.

j. First, when you answer the telephone, remember to identify the company. Not who you are or just say hello.

2. Change the following sentences from passive voice to active voice.
   a. Forty work-hours a week can be saved by the implementation of this new procedure.
   b. A 22% increase in sales is expected this month as a result of the new advertising campaign.
   c. A complete reorganization of the home office administrative staff is being effected by the new chief executive.
   d. The strictest adherence to these rules will be enforced by the security personnel.
   e. It is demanded by the union stewards that this issue of disciplinary authority be immediately brought to the attention of the top management.
   f. The ship was inspected by the skipper.
   g. The dinner was enjoyed by us.
   h. A survey was made by the department of health of the nursing service at the Western State Hospital.
   i. When the play was brought to an end, the actors were greeted with a loud burst of applause by the audience.
   j. Our picture window was punctured by hail, our garage was flattened by winds, and our roof was caved in by a large tree.

3. *Writing a Topic Sentence.* For the following groups of sentences, write a topic sentence that is neither too broad nor too narrow.
   a. Topic _____

   _____

   1. Resorts and hotels at the ocean and in the mountains charge high season rates.
   2. Traffic can sometimes be backed up for five miles.
   3. The heat can make driving very uncomfortable.
   b. Topic _____

   _____

   1. A wide variety of courses is offered at state universities.
   2. Tuition is considerably less than at private institutions.
   3. State universities allow one to interact with people from different ethnic and national groups.

4. Randomly select a paragraph from a textbook. Identify the topic sentence and discuss how the other sentences in the paragraph relate to the topic sentence.

5. Using textbooks, user manuals, or other written material, obtain one example of a logically organized paragraph for each logic sequence: physical or geographic, chronological, emphasis, cause-and-effect, effect-and-cause, and compare–and–contrast.

6. Construct a well organized paragraph on a particular aspect of

one of the following abstract ideas: (a) evaporation of water from a saucer placed in the sun; (b) the killing of elephants for their ivory tusks; (c) the impact of the automobile on U.S. culture.

7. Construct a well organized paragraph on the topic shown using the sequence structure shown:
    a. Physical: flying kite.
    b. Chronological: preparing a speech.
    c. Cause-and-effect: the increase in drug use.
    d. Effect-and-cause: depletion of the ozone layer.
    e. Compare–and–contrast: Shaving with blades versus an electric razor.

8. Write formal sentence definitions of the following terms: cult movie, executive privilege, ghetto, third-world country, glasnost, classic car.

9. Define the following terms using operational or negative definitions: intelligence, a child, love, literacy, substance abuse.

10. Look for poorly written definitions in a professional journal or the newspaper. How could these be improved or clarified?

11. Write a brief definition (formal sentence definition plus two supplementary strategies) for two different audiences or intellectual levels. Which was more difficult to write?

12. Prepare a set of instructions for operating a simple machine. Write a definition for the following terms:

Darcy's law	horsepower
software	viscosity
shear stress	current
buoyancy	acceleration
differential equation	

CHAPTER 4 / **COMMUNICATING TECHNICAL DATA AND STATISTICS**

DO

- learn to properly interpret statistics, not just compute statistical values
- report the units of the mean and standard deviation, not just the numerical value
- report the minimum and maximum values of a data set when characterizing a sample
- report only the significant digits in a final value, but possibly more digits for intermediate values

DON'T

- make bold claims on the basis of a statistical analysis of small samples

4.1 INTRODUCTION

You are hired as a technical expert witness in a legal case between an environmental group and a chemical company. Semiannual measurements of a toxic waste spill are available for the period from 1980 through 1989. How would you present the information to the jury? Plotting the data is one possibility. Fig. 2 shows the concentration as a function of time. Will the plot communicate the same information to the environmentalist as it does to the chemical company executive? Or would each interpret it differently? Would the environmentalist see an increasing trend in the toxin with time? Would the chemical company executive see only the significant scatter and the downward trend from 1987 through 1989? You don't want the jury to make a subjective decision based on emotion. You want to help them make a systematic decision based on a rational analysis of the data.

Consider another case. A union hires you to evaluate the level of

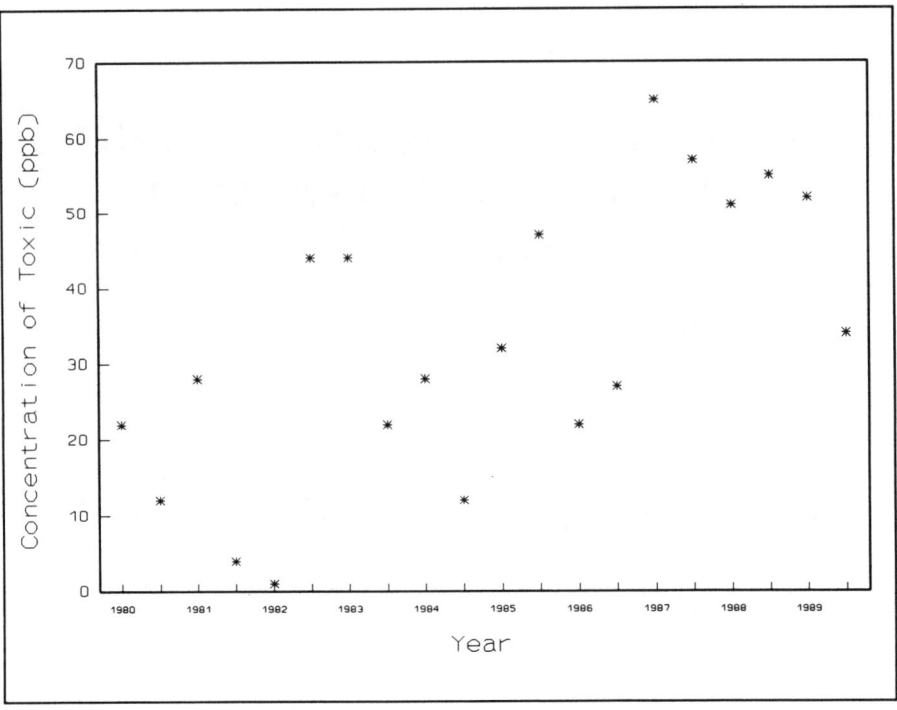

FIG. 2.—Semi-annual Measurements of a Toxic Waste Concentration

noise in a manufacturing plant where the union believes the noise is above
the legal limit. Anything less than 85 decibels (dB) is legally acceptable;
a value of 85 dB or greater is a violation of the limit. You make seven
measurements in the plant, with the following results: 82, 81, 93, 82, 77,
90, and 87 dB. You input these into your computer to compute the mean.
Since your computer prints five digits after the decimal point, the mean
is printed as 84.57143 dB. In your written report to the union, what value
would you show as the mean? 84.57143? 84.6? 85? How would the union
interpret each of these? Do the first two values suggest that the legal limit
is not exceeded? Would the latter value support the union's claim that
the noise level exceeds the allowable limit?

These examples indicate that communicating technical data does have
important consequences. Properly presented data can be the deciding
factor in legal cases and can affect public health. Furthermore, individuals
in technical fields must interpret the data and present the results in a way
that provides the basis for a rational decision. The method of analysis
should be systematic and unbiased. Care must be taken to ensure that
the method of presenting the results communicates the proper results to

those with conflicting interests; it should be presented in a way that minimizes conflict and maximizes the understanding of all stakeholders.

4.2 STATISTICS

Statistics facilitate the communication of technical data and simplify the characteristics of complex information. Statistical methods enable cause-and-effect relationships between variables to be identified. This can reduce conflicts in decision making, thus enhancing communication between the parties involved. Using statistics, observed differences can be tested for significance to determine whether or not they reflect expected variation or out of the ordinary changes.

A purpose of statistics is to provide for a systematic treatment of technical data so that decisions can be agreed upon. This is true whether the material is communicated in a written report or verbally, as in a court of law. A statistical analysis will facilitate decision making based on technical data because it removes some of the opportunity for subjective assessments.

Technical decisions are often based, at least in part, on quantitative data. Very often, the data base is voluminous and must be summarized before it is useful for decision making. Statistical characteristics, which are single-valued indices, enable many numbers to be replaced by one or a few numbers, thus facilitating interpretation. In other cases, decision making is complicated when data are characterized by excessive scatter, leading people to draw different conclusions about relationships between technical variables. The graph of Fig. 2 is an example in which the scatter of the points makes it difficult for people to agree on the presence or significance of a trend in the data. Statistical methods enable relationships to be reduced to a form that is easier to understand. The methods allow two or more people who review both the data and the results of statistical analyses to reach the same assessment.

TABLE 5. Maximum Daily Ozone Concentration (ppb)

Rank	Ozone	Rank	Ozone	Rank	Ozone	Rank	Ozone
1	121	11	79	21	51	31	36
2	109	12	76	22	50	32	36
3	106	13	75	23	46	33	33
4	101	14	71	24	46	34	32
5	97	15	66	25	44	35	30
6	92	16	66	26	43	36	28
7	91	17	63	27	42	37	24
8	91	18	59	28	42	38	23
9	86	19	54	29	39	39	20
10	85	20	53	30	37	40	19

4.2.1   Graphical Analysis of Sample Data

The old saying, "A picture is worth a thousand words," is true in the statistical analysis of data. If the data consist of values of a single random variable, such as the data of Table 5, the data can be reduced using a frequency histogram, which is a figure or tabular summary of the frequency of occurrence in selected intervals of the random variable. A histogram indicates the central tendency of the data, the spread of the data, the presence of extreme events, and the distribution of the data. A moderate sized sample is necessary for a histogram to provide useful information. For small to moderate sized samples, the selection of the bounds of the intervals is important. This can be illustrated using the data of exercise eight at the end of this chapter. The 36 values are used to form three histograms:

Histogram 1		Histogram 2		Histogram 3	
cell	frequency	cell	frequency	cell	frequency
120–124	2	120–129	7	115–124	2
125–129	5	130–139	7	125–134	7
130–134	2	140–149	7	135–144	9
135–139	5	150–159	9	145–154	6
140–144	4	160–169	6	155–164	12
145–149	3				
150–154	3				
155–159	6				
160–164	6				

The first two histograms show a uniform pattern to the ordinates; the third histogram, which appears heavier for the larger speeds, is not uniform, thus suggesting a different distribution of the data. The first histogram, because of the relatively small interval, shows more random variation of the ordinates in comparison with the relatively constant frequency shown by the second histogram. In spite of these problems which are due to the small sample size, the histograms indicate that the speeds are uniformly distributed, with no extreme events and a central tendency at about the center of the histogram (i.e., 145).

4.2.2   Central Tendency

After receiving a graded test, the first piece of information that students want is the class average. Knowing the average gives them a sense of how well they did in comparison to the remainder of the class. The class average is a measure of the central tendency of the grades. The average also measures how easy or difficult the test was.

Three statistics are used as measures of central tendency: the mean, median, and mode. Each has advantages and can communicate different information about the sample from which they were computed.

The mean is a measure of the center, or central tendency, of a sample of data. It is computed as

$$\bar{x} = \frac{1}{n} \sum_{i=1}^{n} x_i \qquad (4\text{-}1)$$

in which $\bar{x}$ is the mean of the set of values x_i and n is the number of values in the set (n is usually called the sample size).

To understand the concept of the mean, consider a seesaw in which cinder blocks are placed along its length, as shown in Fig. 3. At what point should the fulcrum be placed for the seesaw to balance? Since the blocks are not located symmetrically about a point, the solution may not be obvious. If we performed an experiment in which the fulcrum was moved until the seesaw balanced, we would find that a balance was achieved when the fulcrum was at point 9 in Fig. 3. The downward force of the five cinder blocks to the left would just offset the downward force of the five cinder

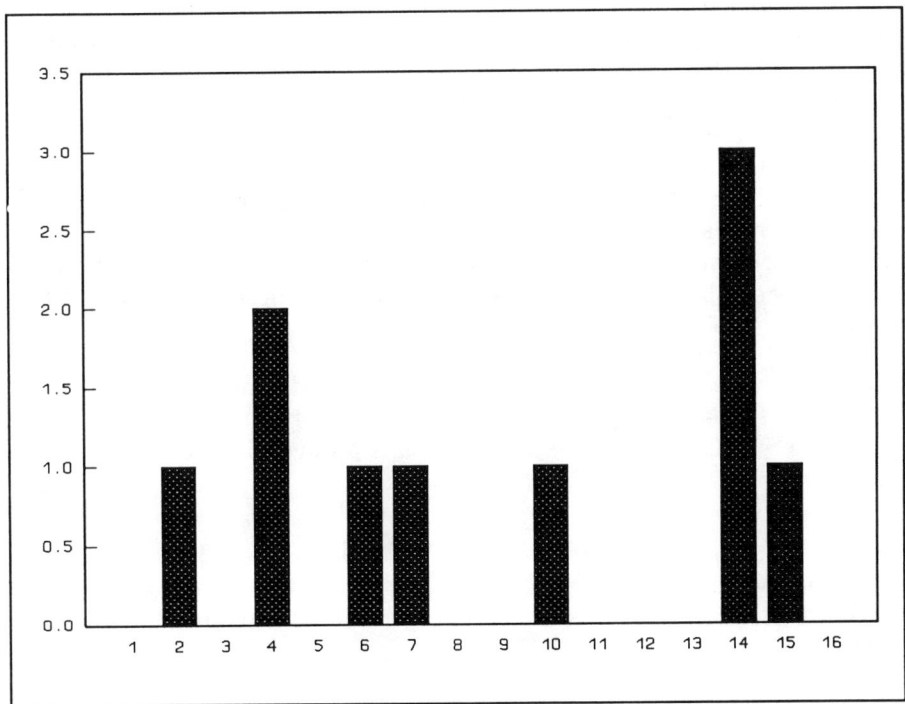

FIG. 3.—*The Mean as the Center of Gravity*

blocks to the right of the fulcrum. It turns out that the mean is a statistical concept that indicates the center of a distribution of data just as the center of gravity is the center of a physical system. In this sense, the mean is a statistical center of gravity.

Maximum daily ozone concentrations for 40 days are given in Table 5, with the data ranked from largest to smallest. The mean of the values $(\bar{x})$ is:

$$\bar{o} = \frac{1}{40} \sum o_i = \frac{1}{40} (2,362) = 59.05 \; ppb \qquad (4\text{-}2)$$

Since the mean has the same units as the random variable, then the mean of the ozone values has units of parts per billion (ppb). What does the mean indicate? It indicates that, on the average, ozone concentrations will be about 59 ppb. But the values of Table 5 indicate that a concentration of 59 ppb occurred on only one of the 40 days, and on many days the concentration was either much larger or much smaller than 59 ppb. The mean might be useful if it were compared to some standard, such as the ozone concentration that represented a health hazard. Then the sample mean would suggest whether or not there was, on the average, a problem.

In other cases, the comparison of two means may communicate important information. The means of the pollutant concentrations of Fig. 2 for the 1980–84 and 1985–89 periods are 21.2 ppb and 43.64 ppb, respectively. This suggests that, on the average, the toxic concentration increased by 106%. The decision maker must decide whether or not this is a significant increase in central tendency.

The median is another useful measure of central tendency, especially for small samples. When determining the median, the data must be ranked from largest to smallest, with a rank of 1 for the largest value and a rank of n for the smallest value in the sample. Mathematically, the median (m) is:

$$m = \begin{cases} x_j & \text{when } n \text{ is an odd integer} \qquad (4\text{-}3a) \\ \dfrac{x_k + x_{k+1}}{2} & \text{when } n \text{ is an even integer} \end{cases}$$

$$(4\text{-}3b)$$

with $j = (n + 1)/2$ and $k = n/2$.

For the 40 ozone values of Table 5, the median is computed with Eq. 4-3b:

$$m_o = \frac{x_{20} + x_{21}}{2} = \frac{53 + 51}{2} = 52 \; ppb \qquad (4\text{-}4)$$

where $k = 20$. In this case the mean is about 13% larger than the median because there are a few very large ozone measurements in the sample; thus, the sample is skewed toward the high values. If the sample had approximately an equal proportion of high and low values, then the mean and median would be closer to each other.

For small samples that contain an extreme value, the mean can be a misleading measure of central tendency. Consider the sample of 5 measurements: 0.012, 0.037, 0.203, 0.546, and 96.4. The mean of the sample is 19.44, which is considerably larger than four of the five values. The median is $x_3 = 0.203$ and may be a better reflection of the "typical" value of x.

The mode is a third measure of central tendency. The mode of a group of data is the value that occurs most frequently. For example, in the sample 2,8,9,2,4,2,2,3,8,7, the mode is 2. In the sample 2,8,8,9,2,4,8,2,3,7, there are two modes since both 2 and 8 occur three times; this is called a bimodal sample. For the sample 2,8,9,4,3,6,7,10, there is no mode since each value occurs the same number of times.

4.2.3   Standard Deviation

While students are always interested in the average of the grades, they should also be interested in the dispersion of the grades. For example, if Kaye gets a 60 on a test that had a mean of 50, she will be happier if the grades ranged from 40 to 60 than she would be if the grades ranged from 10 to 90. In the first case, her grade was the highest; in the second case, it was only slightly above the mean. Thus, if Kaye wants to make a decision on how well she did on the test, she needs more information than the mean provides; she must know something about the dispersion of the data, i.e., how the data vary about the mean.

The standard deviation, which is the most useful measure of dispersion in a data set, is computed by

$$S = \left\{ \frac{1}{n-1} \sum_{i=1}^{n} (x_i - \bar{x})^2 \right\}^{0.5} \tag{4-5a}$$

$$= \left[\frac{1}{n-1} \left(\sum_{i=1}^{n} x_i^2 - \frac{1}{n} \left(\sum_{i=1}^{n} x_i \right)^2 \right) \right]^{0.5} \tag{4-5b}$$

in which S is the standard deviation. Equation 4-5a indicates that the standard deviation is a measure of the deviation of values about the mean. Equation 4-5b is more useful for computation since it does not require the prior calculation of the mean.

The sample of five measurements given in Table 6 can be used to

TABLE 6. Calculation of the Sample Standard Deviation

	x	$x - \bar{x}$	$(x - \bar{x})^2$	x^2
	12	−10	100	144
	18	−4	16	324
	22	0	0	484
	23	1	1	529
	35	13	169	1225
Sums	110	0	286	2706

compute the standard deviation with Eqs. 4-5. For the mean square calculation of Eq. 4-5a

$$S = \left[\frac{1}{5-1} (286) \right]^{0.5} = 8.46 \tag{4-6}$$

and for the computational formula of Eq. 4-5b

$$S = \left[\frac{1}{5-1} \left(2706 - \frac{1}{5} (110)^2 \right) \right]^{0.5} = 8.46 \tag{4-7}$$

For this data set, three of the five values of x lie within the range from $\bar{x} - S$ ($= 13.54$) and $\bar{x} + S$ ($= 30.46$), which indicates that the standard deviation reflects the spread of the data.

Previously, the question was asked, Is the difference between the 1980–84 and 1985–89 means of Fig. 2 important? The standard deviations of the measurements may provide some insight into the question. The standard deviations are 14.1 ppb and 14.2 ppb for the two periods, respectively. These indicate that there is considerable scatter in the two samples and the difference in means of 22.4 ppb seems less important than was suggested by the change in means of 106%. Thus, the standard deviations have enabled us to better interpret the difference in means.

Just as the mean reflects the center of gravity of a mass, the standard deviation has an analogy in mechanics. The inertia of a mass is a measure of the force that would be necessary in order to accelerate it. For example, when you swing a stone attached to a string in a circular path, you must exert a continual force on the string. The inertia of the stone resists the continual change of direction. The variance, which is the square of the standard deviation (S^2), is the moment of inertia of a distribution of mass. Just as the mean corresponds to the center of gravity, the standard deviation corresponds to the radius of gyration.

The standard deviation of the locations of the cinder blocks on the seesaw of Fig. 3 equals 5.0. If the seesaw were allowed to rotate about the mean (9), then the standard deviation reflects the distance from the mean where a force would have to be applied to keep the seesaw from rotating. If five cinder blocks were located at point 3 and the other five blocks at point 15, the standard deviation would be 6.3. This larger radius of gyration reflects the fact that there are no values close to the center of gravity.

4.2.4 Data Extremes

When interpreting technical data, it is always important to examine the extremes of the sample. If a data set contains an extremely large or extremely small value, it should be discussed in the written report. Extreme values may be the result of measurement or computation error, an extreme observation in a small sample, or an indication of a mixed population. In discussing the interpretation of the mean, a sample of five was given: 0.012, 0.037, 0.203, 0.546, 96.4. In this case, the largest event is about 175 times larger than the next largest value. The value of 96.4 may be a legitimate value, but it appears as an extreme event only because of the small sample. As another example, consider the case of a very large flood caused by a hurricane. If just one hurricane-generated flood occurred in a 25-year record of floods and it was much larger than all the other floods, then the extreme flood would be the result of a mixed population of flooding, hurricane and non-hurricane floods.

The range of the data is the difference or interval between the minimum and maximum of the sample. When using sample data to draw more general inferences, analysis beyond the range of the data represents extrapolation and the accuracy of decisions beyond the range of the sample data is assumed to be less than accurate within the range of the data. The extreme values of a sample, as well as the mean and standard deviation, should always be reported when summarizing an analysis of technical data. Otherwise, a user of the report may apply the results well beyond the range of the data, thus assuming a false sense of accuracy.

In reviewing the ozone measurements of Table 5, the minimum and maximum values are 19 and 121 ppb, respectively. The minimum is not extremely low since there are three other values below 25 ppb. While the second largest value is 12 ppb lower than the maximum value, the maximum is not considered an outlier or unexpected value since it is only about 10% larger than the second largest recorded value. When reporting such data, it is important to evaluate the measurements at the extremes to ensure that the values are reasonable measurements. If the largest value had been an order of magnitude larger than the second largest value, say 1,000 ppb, then an explanation would be required to justify using such as extreme event in calculations based on the sample.

In one case, a researcher presented an equation for predicting the concentration of a pollutant as a function of land area. The equation was developed using land areas from 50 to 150 acres, but the researcher failed to report these minimum and maximum values. Someone tried to use the equation, which was reported in a professional journal, for a land area of 60 square miles, which is over 250 times larger than the maximum land area used to develop the equation. For 60 square miles, the equation yielded a predicted value of 2 million ppm, which is physically impossible. It is fortunate that the predicted value was so extremely poor; otherwise, the user may have obtained and used a value that was physically possible, but highly inaccurate.

4.2.5 Normal Distribution

When plotting a histogram of sample data, such as the grades on a test, the data often appear as a bell-shaped curve or distribution, with many values near the mean but a few values at the extremes. It is often assumed that such data have a normal distribution, which is a commonly used probability function. The normal distribution is popular because the histograms of many data sets have the characteristic bell-shaped form and much of statistical theory assumes that the data are normally distributed. If a histogram of the logarithms of the data plot with a bell-shaped curve, then the random variable may have a log-normal distribution, which means that the logarithms are normally distributed.

To compute normal probabilities for a random variable x, which has mean $\bar{x}$ and standard deviation S, the following transformation to a new random variable z can be made:

$$z = \frac{x - \bar{x}}{S} \tag{4-8a}$$

where z has a normal distribution with a mean of 0 and a standard deviation of 1. Equation 4-8a can be rearranged to compute the value of x for a given value of z:

$$x = \bar{x} + zS \tag{4-8b}$$

Values of the cumulative normal distribution $p(z < z_0)$ are given in Table 7; more complete tables can be found in statistics books. To find the probability $P(z > z_0)$, the value of Table 7 can be subtracted from 1: $P(z > z_0) = 1 - P(z < z_0)$.

If a histogram of sampled data appears bell shaped, then the probability of any value of the random variable being equal or exceeded can be obtained using Eq. 4-8a and Table 8-3. Fig. 4 shows a frequency histogram of the base 10 logarithms for 58 years of annual maximum dis-

TABLE 7. Values for $p(z < z_0)$ of the Cumulative Normal Probability Function for Selected Values of the Standardized Variate z

z	p	z	p	z	p	z	p
−3.0	0.001	−0.60	0.274	0.05	0.520	0.7	0.758
−2.5	0.006	−0.50	0.309	0.10	0.540	0.8	0.788
−2.0	0.023	−0.45	0.326	0.15	0.560	0.9	0.816
−1.8	0.036	−0.40	0.345	0.20	0.579	1.0	0.841
−1.6	0.055	−0.35	0.363	0.25	0.599	1.2	0.885
−1.4	0.081	−0.30	0.382	0.30	0.618	1.4	0.919
−1.2	0.115	−0.25	0.401	0.35	0.637	1.6	0.945
−1.0	0.159	−0.20	0.421	0.40	0.655	1.8	0.964
−0.9	0.184	−0.15	0.440	0.45	0.674	2.0	0.977
−0.8	0.212	−0.10	0.460	0.50	0.691	2.5	0.994
−0.7	0.242	−0.05	0.480	0.60	0.726	3.0	0.999
		0.00	0.500				

charge data, which are given in exercise 4-11. The mean and standard deviation of the logarithms are 3.889 and 0.2031, respectively. While the data appear slightly skewed to the left, a normal distribution can be assumed for the distribution of the logarithms, that is the discharges are log-normally distributed. The log-normal distribution can be used to compute probabilities of a specific discharge being exceeded. For example, the probability of a discharge of 20,000 cfs being exceeded in any year is:

$$P\,(x > 20{,}000) = P\left(z > \frac{\log(20000) - 3.889}{0.2031}\right) = P\,(z > 2.03)$$

$$= 1 - P\,(z < 2.03) = 1 - 0.979 = 0.021$$

This example illustrates how the histogram can be used to hypothesize a probability distribution, such as the normal curve, from which probability estimates can be made. It would be difficult to make probability estimates without making an assumption of the underlying probability function.

4.2.6 Regression and Correlation Analysis

In the introduction, data of a toxic waste versus time were introduced (Fig. 2). How can the data be presented to reflect the processes that were responsible for the measured data? The environmentalist suspects that there is a trend in the data, but knows that a subjective argument that the trend exists will not hold up in court. The environmentalist needs a systematic theoretical analysis of the data to support the claim of a trend.

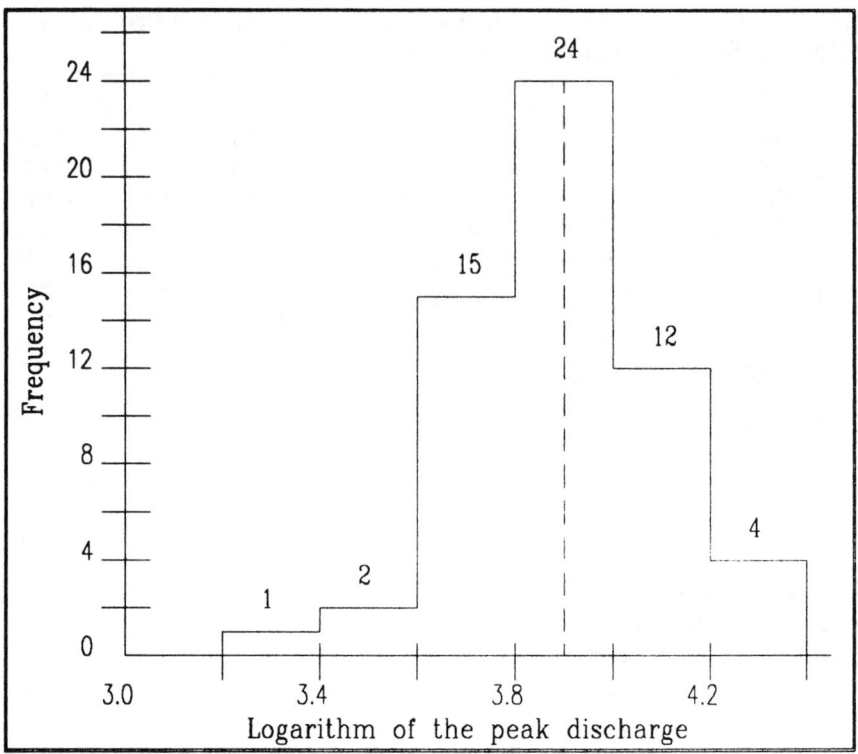

FIG. 4.—*Histogram of the Logarithms of the Annual Peak Discharge Record for the Piscataquis River at Dover-Foxcroft, Maine*

Linear trends can be fit using regression analysis. Given a set of n pairs of measurements on the independent, x, and dependent, y, variables, the following linear equation can be fit:

$$\hat{y} = b_o + b_1 x \qquad (4\text{-}9)$$

in which $\hat{y}$ is the predicted value of y, b_o is the intercept coefficient, and b_1 is the slope coefficient. Values can be computed for b_o and b_1 using the method of least squares, which involves minimizing the sum of the squares of the deviations, that is minimize

$$\sum_{i=1}^{n} (\hat{y}_i - y_i)^2 \qquad (4\text{-}10)$$

A least squares analysis yields the estimators b_1 and b_o:

$$b_1 = \frac{\Sigma xy - (\Sigma x \Sigma y)/n}{\Sigma x^2 - (\Sigma x)^2/n} \tag{4-11a}$$

$$b_o = \Sigma y/n - b_1 \Sigma x/n \tag{4-11b}$$

in which each summation is over all n observations.

The accuracy of predictions made using Eqs. 4-9 and 4-11 can be assessed using two goodness-of-fit statistics: the correlation coefficient, R, and the standard error of estimate, S_e:

$$R = \frac{\Sigma xy - (\Sigma x \Sigma y)/n}{[\Sigma x^2 - (\Sigma x)^2/n]^{0.5} [\Sigma y^2 - (\Sigma y)^2/n]^{0.5}} \tag{4-12}$$

$$S_e = \left[\frac{1}{n-2} \Sigma (\hat{y} - y)^2 \right]^{0.5} \tag{4-13}$$

The value of S_e is frequently expressed as a ratio, S_e/S_y, where S_y is the standard deviation of the measured values of y. A value of R of 0 means that there is no linear association between y and x, which implies that variation in y is not *caused* by x. A value of R of 1 means that the linear model provides perfect estimates of y, which implies a causal relationship between y and x. When n is small, high values of R are expected even when the causal relationship is not good. The square of the correlation coefficient, R^2, represents the fraction of variation explained by the regression equation. For small samples, the S_e/S_y ratio may be a better indicator of prediction accuracy than R. The smaller the value of S_e/S_y, the more accurate the estimates of y can be expected.

Using the method of least squares, the environmentalist produces the following regression equation (see Fig. 5):

$$\hat{y} = 11.587 + 1.9860 \, x \tag{4-14}$$

in which $\hat{y}$ is the predicted value of the pollutant and x is the observation indicator ($x = 1$ for first measurement in 1980, $x = 2$ for the second measurement in 1980, and $x = 20$ for the second measurement in 1989). The environmentalist points out that $\hat{y}$ equals 51.3 ppb for $x = 20$ and $\hat{y}$ equals 13.6 for $x = 1$; thus, the equation indicates that the pollutant concentration increased by 277% during the decade. Is the environmentalist justified in concluding that the increase is significant?

The chemical company executive, who is familiar with some statistical methods, computes a standard error of estimate of 13.92 ppb. In comparing this to the standard deviation of the pollution measurements, S_y, the S_e/S_y ratio is 0.776, which means that the errors are still relatively large. The executive also computes the correlation coefficient ($R = 0.655$)

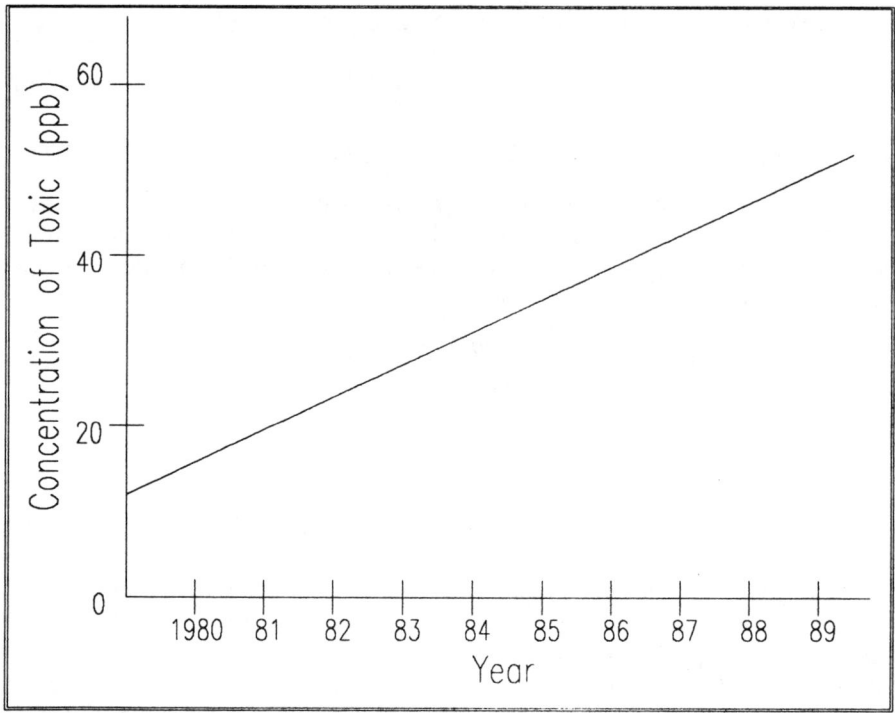

FIG. 5.—*Regression of Toxic Concentration on Time*

and finds that it is not significantly different from a correlation of 0 when the level of significance of 0.1% is used. Futhermore, R^2 equals 0.43; the chemical company executive argues that the linear regression line of Eq. 4-14 explains less than 50% of the total variation, that is the error variation is greater than the explained variation. Is the chemical company executive justified in concluding that the increase is not significant when the small sample ($n = 20$) is taken into consideration?

You are the expert witness and based on your experience and knowledge you conclude that both the environmentalist and the chemical company executive have correctly computed the regression equation, the standard error ratio, and the test of significance on the correlation coefficient. But which one has made the correct interpretation of the technical data?

Proper interpretation of technical data can have important consequences. If the environmentalist's interpretation is adopted, the chemical plant may have to close, which means the loss of jobs and the subsequent loss in taxes for the local economy; additionally a costly clean-up may be ordered, which means that other social services may have to be delayed. Conversely, if the chemical company executive's interpretation is accepted, then it is possible that the chemical plant will continue to pollute the groundwater. It is a decision that does not have a certain answer; the uncertainty in the results of the data analyses is quite significant but a decision must be made.

4.2.7 Sample Size

As indicated above, the sample size is a measure of accuracy. Theory shows that the accuracy of a statistic computed from sample data usually improves as the sample size increases. For small samples, the accuracy is relatively poor and almost any finding could result. The sample size should always be reported when summarizing data; it can be just as important as the mean or standard deviation.

As an extreme example, consider the statistical significance of a sample correlation coefficient (R) of 0.87 that is based on a data set for a sample of 5. Using a statistical hypothesis test, it can be shown that a sample R of 0.87 is not statistically different from 0 and from 0.99, which implies that there is no association between the two variables ($R = 0$) and that there is almost perfect association ($R = 0.99$). Clearly, with a small sample, we can prove that almost anything is true. Therefore, when communicating technical data, it is important to report the sample size because it indicates the degree of confidence that we will have in decisions based on the data.

4.3 SIGNIFICANT DIGITS

Consider the problem of measuring the distance between two points using a ruler that has a scale with 1 mm between the finest divisions. If we record our measurements in centimeters and if we estimate fractions of a millimeter, then a distance recorded as 3.76 cm gives two precise digits (i.e., the 3 and the 7) and one estimated digit (i.e., the 6). If we define a significant digit to be any number that is relatively precise, then the measurement of 3.76 cm has three significant digits. Even though the last digit could be a 5 or a 7, it still provides some information about the length and so it is assumed to be significant. If we recorded the number as 3.762, we would still have only three significant digits because the last digit (i.e., the 2) is not trustworthy since the 6 is not precise. Only one imprecise digit can be considered as a significant digit.

The number of significant digits is used to reflect the accuracy of the number. For example, if the number 46.23 is exact to the four digits shown, it is said to have four significant digits. The error is then no more than 0.005.

The number of significant digits is set either by the scale of measurement or the physical significance of the numbers. For example, a digital bathroom scale that shows weight to the nearest pound uses up to three significant digits. If the scale shows, for example, 159 pounds, then the individual assumes his or her weight is within 0.5 pounds of the observed value. In this case, the scale has set the number of significant digits. The number of significant digits in someone's height is set by the physical significance of the numbers. The ruler may provide for measurements to ¹⁄₃₂ of an inch but the number may be recorded to the nearest inch because more precise values are not important. In that case we would have two significant digits, since a measurement of 5-ft, 11³⁄₃₂ in. would be recorded as 71 in. In this case, the meaningfulness of the information content of the number set the number of significant digits.

In Section 4.1, seven measurements of the noise dose were given; each was recorded to the nearest integer value, and we will assume that the smallest increment of scale on the measuring instrument was one decibel. So what is the mean of the seven measurements? Strictly speaking, since the measurements were made to two significant digits, the computed value is only accurate to two significant digits. In the report written for the union, only two significant digits should be specified. Thus, the mean is 85 dB, which equals the threshold of unacceptable noise.

The rule for identifying the number of significant digits is: The digits from 1 to 9 are always significant, with zero being significant when it is not being used to set the position of the decimal point. For example, each of the following have three significant digits: 2,410, 2.41, and 0.00241. In the first example, the 0 is only used to set the decimal place. Confusion can be avoided by using scientific notation: 2.41×10^3. This means that 2.41×10^3 has three significant digits, where 2.410×10^3 has four significant digits. The numbers 13 and 13.00 differ in that the former is recorded at two significant digits, and the latter has four significant digits. In this case, the two zeros are significant.

The location of the decimal point does not influence the number of significant digits. The measurement of 3.76 cm could be written as 37.6 mm or 0.0376 m; each of these three values has three significant digits.

When performing computations, the general rule on setting the number of significant digits in a computed value is: Any mathematical operation using an imprecise digit is imprecise. When combined with the rule that only one imprecise digit can be considered significant, then the number of significant digits is set. Consider the following multiplication of two numbers, each having three significant digits, with the last digit of each being imprecise:

$$
\begin{array}{r}
4.26 \\
\underline{8.39} \\
\underline{3834} \\
127\underline{8} \\
340\underline{8} \\
\hline
35.\underline{7414}
\end{array}
$$

The digits that depend on imprecise digits are underlined. In the final answer, only the first two digits (35) are not based on imprecise digits. Since one and only one imprecise digit can be considered as significant, then the result is recorded as 35.7.

Just because a set of numbers is only accurate to a specified number of digits does not mean that computations based on those numbers should be rounded and reported to that number of digits. Computed values should only be rounded at the end of the computation process. For intermediate steps, the computed numbers should be computed and reported to more digits than the number of significant digits in the original set of numbers. For example, the regression equation for the data of Fig. 2, with the coefficients reported to five digits, is shown as Eq. 4-14. For three digits the equation would be:

$$\hat{y} = 11.6 + 1.99 \, X \tag{4-15}$$

Using three, four, and five digits for the regression coefficients yields the following predicted values ($\hat{y}_3$, $\hat{y}_4$, and $\hat{y}_5$) for given values of x:

x	$\hat{y}_3$	$\hat{y}_4$	$\hat{y}_5$
1	13.59	13.576	13.573
20	51.40	51.310	51.307
40	91.20	91.030	91.027
100	210.60	210.19	210.187

Rounding the regression coefficients to three digits does not cause an error for $x = 1$, but errors of 0.1, 0.2, and 1 occur for $x = 20$, $x = 40$, and $x = 100$, respectively, assuming that the final values are rounded to three significant digits. Using the regression coefficients with four digits produces estimated values for all of the values of x that are without rounding error when evaluated at three significant digits. In this case, estimated values of y can be rounded to three significant digits, but the regression coefficients should be not rounded to three digits; the rounding should be made at the end of the computation, not at intermediate calculations.

4.4 ERRORS

If you take your body temperature with a standard mercury thermometer, how accurate is the thermometer reading? If 10 measurements were made with the thermometer, what would be the range of the readings? If the scale on the thermometer was inaccurately printed such that all readings were in error by 0.3°F, is this a significant error?

While these may seem like trivial questions, their underlying basis relates to many professional problems that involve technical data. What

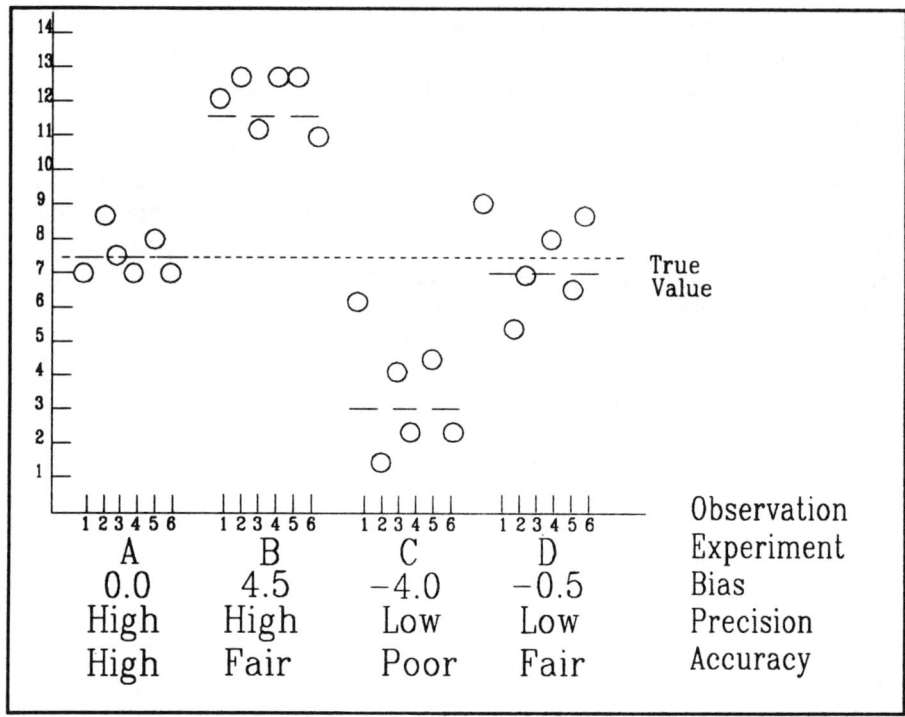

FIG. 6.—*Assessment of Bias, Precision, and Accuracy of Six Measurements from Each of Four Laboratories*

if a timer for the space-shuttle booster rockets were miscalibrated and caused the fuel to shut down 20 seconds before it was supposed to? What if a laboratory instrument that measures the compressive strength of concrete consistently indicates that the strength is 10% higher than it really is? These errors can have very significant consequences. Therefore, it is important to indicate the expected accuracy of an instrument when communicating technical data.

An error is the difference between the measured value and the true value. Errors can be classified as being either systematic or random. If repeated measurements are made with the same instrument and the average of the repeated measurements is not close to the true value, then the instrument causes a systematic error, which is the difference between the mean of the measurements and the true value. This difference is called the bias. A consistent overprediction represents a positive bias; a consistent underprediction is a negative bias. It may be possible to eliminate the bias by calibrating the instrument.

If repeated measurements are made and they differ from the true value, but the mean of the measurements is not significantly different from the true value, then the individual errors are said to be random errors. The variation of the random errors is a measure of the precision of the measuring instrument. If the random errors are relatively small, the instrument is said to be precise. If the errors are large, then the instrument gives imprecise measurements.

The accuracy of a set of measurements is influenced by both the bias and precision. Inaccurate measurements can result from either a bias in the instrument or a lack of precision, or both.

Fig. 6 shows six measurements from each of four experiments. In experiment A, the sample mean equals the true mean so the measurements are unbiased; the small variation of the measurements about the sample mean indicates that the measurements are precise. Since there is high precision and no bias, the method of measurement has relatively high accuracy.

In experiment B, the measurements are positively biased since the sample mean is greater than the true mean; however, the measurements are fairly precise because they bunch around the sample mean. So in spite of good precision, the accuracy would only be fair because of the poor bias.

In experiment C, both the bias and precision are poor, which means the accuracy is poor. The bias is negative, which means that the measurements are consistently less than the true value.

In experiment D, the precision is low but the bias is small; therefore, the accuracy is fair.

In communicating technical data, every effort must be made to explain errors in measurements. The task is difficult because it is rarely possible to know the true value. In such cases, the bias is unknown, so the accuracy cannot be assessed; it can only be estimated. The bias can be assessed if an assumption is made that some standard of comparison is used in place of the true value. For example, in regression analysis, the regression line is used as the standard of comparison and the errors can be computed. Unless all of the errors are negligibly small or zero, they contain information that should be explained. It is inadequate to show the regression equation and not evaluate the residuals. At the minimum, the residuals can be characterized by their standard deviation, which is the standard error of estimate of Eq. 4-13. The cause of extremely large residuals, whether positive or negative, should be evaluated and explained in the report. The report should also include a discussion of residuals that show correlation between themselves, such as when all the residuals in one area of the plot are of the same sign, which indicates a local bias. The important point is: While a regression equation provides an explanation

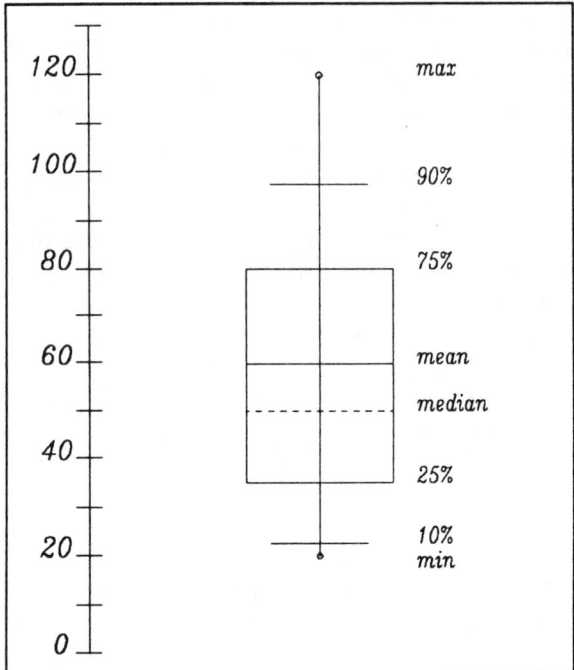

FIG. 7.—*Box—and—whisker Plot for Maximum*
Daily Ozone Concentration (ppb)

of the relationship between two variables, a considerable amount of information may be in the residuals that needs to be explained, whether it reflects a bias or a lack of precision.

4.5   BOX-AND-WHISKER PLOTS

Box-and-whisker plots are a graphical method for showing the distribution of sampled data, including the central tendency (mean and median), dispersion (10, 25, 75, and 90 percentiles), and the extremes (minimum and maximum). Additionally, they can be used to show the bias about the standard value and the relative sample size, if the figure includes multiple plots for comparison.

To construct a box-and-whisker plot, compute the following characteristics of a data set:

1. The mean and median of the sample.
2. The minimum and maximum of the sample.
3. The values that 90, 75, 25, and 10% of the sample are less than or equal to.

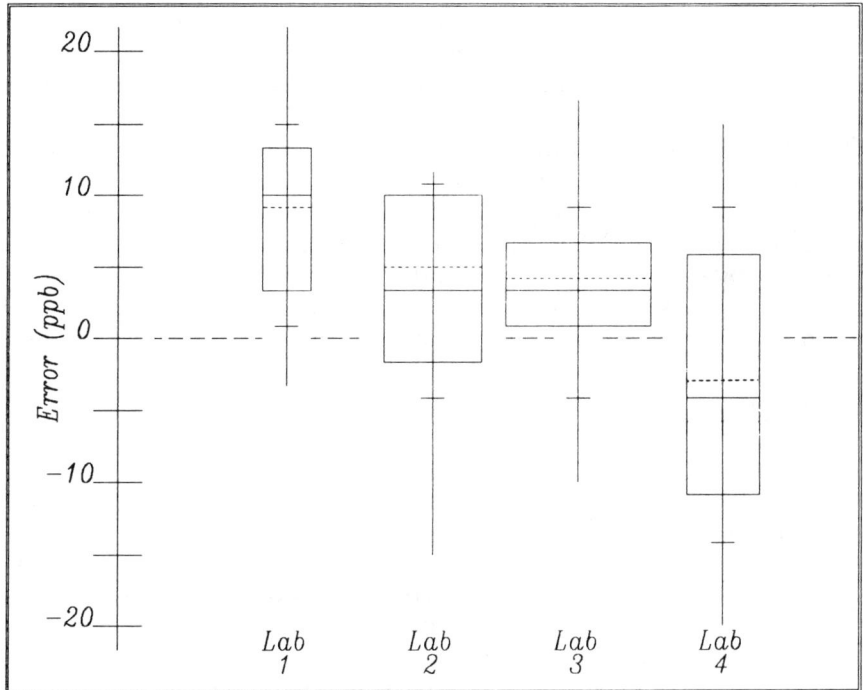

FIG. 8.—*Display of Multiple Box–and–whisker Plots*

The plot consists of a box, the upper and lower boundaries that define the 75 and 25 percentiles, and upper and lower whiskers, which extend from the ends of the box to the extremes (see Fig. 7). At the 90 and 10 percentiles, bars, which are one-half of the width of the box, are placed perpendicular to the whiskers. The mean and median are indicated by solid and dashed lines that are the full width of the box.

Fig. 7 shows the box-and-whisker plot for the 40 observations of the maximum daily ozone concentration given in Table 5. The mean and median are 59 and 52 ppb, respectively. The 10, 25, 75, and 90 percentile points are 24, 36, 79, and 97 ppb, respectively.

If a figure includes more than one box-and-whisker plot and the samples from which each plot is derived are of different sample sizes, then the width of the box can be used to indicate the sample size, with the width of the box increasing as the sample size increases. Fig. 8 shows box-and-whisker plots of samples of a toxic chemical analyzed at four different laboratories. Since the samples were of different concentrations, the distributions are presented as the differences between the true concentration and the concentration reported by the laboratory, that is the error. Laboratories 1, 2, 3, and 4 processed 12, 29, 38, and 19 samples, respectively. The widths of the plots reflect the different sample sizes. Greater accuracy can be expected for the statistics based on the larger sample sizes.

Box-and-whisker plots can also be used to assess sample biases, when the true values are known. In the case of Fig. 8, the true concentrations of the samples containing the pollution were known; therefore, the mean errors are an indication of the laboratory bias. Laboratories 1, 2, and 3 have positive biases of about 9.5, 2.0, and 2.5 ppb, respectively. Laboratory 4 has a negative bias of 3.5 ppb. If standard deviations of the sample data were computed, the significance of the biases could be tested statistically.

4.6 EXERCISES

1. Find three histograms in new magazines or technical journals. Are the scales of the ordinate chosen so that the authors can make a particular point?
2. Compose a rule for selecting the width of the intervals for a histogram. In one paragraph, write a justification of the rule, with the paragraph, write a justification of the rule, with the paragraph intended for a technical audience.
3. Compute the mean and standard deviation of the data for exercise 4-7.
4. For the data of exercise 4-13 compare the mean and median. Discuss the importance of the difference.
5. The mean corrosion penetration for five specimens of a particular type of steel is 120 mils. In one paragraph, explain the meaning of the statistic to a nontechnical audience. Would knowledge of the standard deviation increase or decrease the complexity of the explanation? Explain.
6. Using samples of data, the mean time until a pothole develops is computed for local roads (7.2 years), state roads (9.6 years), and interstate highways (12.5 years). In one paragraph, discuss these statistics in terms that nonengineer would understand. Would the explanation differ if the means were based on a sample of 5 rather than 500? Explain.
7. In one paragraph, explain to a nontechnical person the importance of the standard deviation in making technical decisions.
8. If we assume the winning times for the Indianapolis 500 (see exercise 4-13) are normally distributed, compute the probability that the speed will be greater than 160 mph.
9. If we assume the data of exercise 4-13 have a log-normal distribution (log mean = 3.889, log standard deviation = 0.2031), find

the probability that the flow rate is less than 2500 cfs. Does the log-normal probability agree with the sample probability?

10. Would it be appropriate to fit a linear regression line to the data of Fig. 25 showing bike helmet use in Texas? Explain.

11. Using the data of Fig. 22, regress the discharge (y) on the year (X). Compute the correlation coefficient and S_e/S_y ratio. Discuss the goodness of fit and the appropriateness of the linear model.

12. An engineer regresses the proportion of the total project cost (Y) expended upon completion of the proportion of the project duration (X), with the resulting model:

$$Y = 0.03 + 0.95 \, X$$

13. The following are the winning speeds (mph) for the winners of the Indianapolis 500 from 1949 through 1984. Has the increase in winning times changed significantly during the period?

1949	121	1958	134	1967	151	1976	149
50	124	59	136	68	153	77	161
51	126	60	139	69	157	78	161
52	129	61	139	70	156	79	159
53	129	62	140	71	158	80	143
54	131	63	143	72	163	81	139
55	128	64	147	73	159	82	162
56	128	65	151	74	159	83	162
57	136	66	144	75	149	84	164

14. The following are the surface area, A (sq. mi), and depth, d (ft), of world lakes. Is there a significant relationship between A and d?

Lake	A	d	Lake	A	d
Caspian Sea	143,244	3,363	Malawi	11,150	2,280
Victoria	26,828	270	Maracaibo	5,270	115
Erie	9,910	210	Nicaragua	3,100	230
Aral Sea	24,904	220	Manitoba	1,799	12
Chad	6,300	24	Balkhash	7,115	85

15. Obtain a set of data from the newspaper or a consumers magazine. Compute the mean and standard deviation and provide an interpretation of the results. Also plot a histogram or box-and-whisker plot and discuss the results.

16. Develop a box-and-whisker plot of the following flood discharges (cubic feet per second):

21,500	12,900	8,690	7,420	6,240	4,680
19,300	11,600	8,600	7,380	6,200	4,570
17,400	11,100	8,350	7,190	6,100	4,110
17,400	10,400	8,110	7,190	5,960	4,010
15,200	10,400	8,040	7,130	5,590	4,010
14,600	10,100	8,040	6,970	5,300	3,100
13,700	9,640	8,040	6,930	5,250	2,990
13,500	9,560	8,040	6,870	5,150	2,140
13,300	9,310	7,780	6,750	5,140	
13,200	8,850	7,600	6,350	4,710	

17. Find an issue of a weekly news magazine (e.g., *Time*, *Newsweek*). How are data presented? Compare this presentation to those in a textbook or technical journal. What are the differences?

CHAPTER 5 / **ETHICAL CONSIDERATIONS IN COMMUNICATION**

DO

- reference ideas, not just direct quotes
- learn proper referencing formats
- obtain written permission when publishing copyrighted material
- use gender-neutral writing

DON'T

- paraphrase someone's writings without properly referencing the source
- use libelous writing

5.1   INTRODUCTION

Respect is an important human value, whether for oneself, others or the environment. With respect to communication, respect refers to the consideration or honor one gives to the works of others. Just as we should respect the property of our neighbors, it is important to respect the effort that others have expended in their work. We honor this work by giving proper credit, usually in the form of referencing. Respect forms the basis for much of what is presented in this chapter.

Students commonly ask: When should I reference the work of others? If we had to reference only direct quotes, the solution would be simple and obvious. But credit via a proper reference should also be given whenever the idea of another is used, even if it is a loose summary or paraphrase. Failure to properly credit the work of others constitutes plagiarism, which is a very serious form of misconduct. It is not, however, the only form. In some cases, referencing is not sufficient; permission to reprint copyrighted material may be required.

In addition to plagiarism, other aspects of writing present ethical dilemmas. Certainly, writings should be free of sexism, racism, ethnocentrism, and bias. Libel is another problem; while it is not a major concern in technical communication, it must be considered by all writers.

This chapter focuses on aspects of communication that require ethical considerations. These include referencing and plagiarism, the elimination of sexism and bias in writings, libel, and requesting permission to reproduce copyrighted material.

5.2   DIRECT QUOTATIONS AND PARAPHRASING

In professional communications, such as theses, project reports, or professional journal publications, it is necessary to give credit for direct quotations through referencing (see Chapter 6). A proper reference is adequate credit in most cases; it represents fair use of the material and meets the guidelines of the most recent copyright law. However, under certain circumstances, referencing is not sufficient, and written permission is required. Unfortunately, the copyright laws do not give a specific lists of do's and don'ts for guidance. When in doubt, it is best to seek written permission prior to publishing the material.

Written permission should be obtained for a quotation that is long (500 words or an entire paragraph), for a partially or completely reproduced table or figure, or any poetry or music (even one line). Permission is required for printing statistical tables that are copyrighted. Permission should be requested from the holder of the copyright. If the material is not copyrighted, then permission should be obtained from the author. A copy of the material in the form in which it will be published should be included with the request.

While direct quotations must be referenced, it is also necessary to reference paraphrased material. Paraphrasing is the restatement of an idea in words different from those used by the original author. Even the use of the organizational structure of written material constitutes paraphrasing. Again, copyright laws are not specific, so written permission should be requested when the amount of paraphrased material is similar to that for direct quotations.

The comments in this section do not just apply to the practicing engineer and engineering researcher; they also apply to students. Is it any different for a student to use the structure and wording from a classmates laboratory report than it is for the researcher to use a technical paper or report of another professional? In both cases, such use or misuse, represents plagiarism. In both cases, it is necessary for the author to properly reference the previous work of others. We recognize it is wrong to copy the work of others during a test; it is also wrong to copy someone else's report, even the structure of a report. Placing one's name on the coversheet of the laboratory report implies that the report was completed using only the works that are cited in the report. Would a student who uses a fellow student's report without giving proper credit be likely to withhold appropriate credit from a research report written years after his or her graduation? Sometimes unethical conduct practiced in college carries over to professional life.

5.3 GENDER NEUTRAL WRITING

Examples of biased writing are common in books and articles that were published in the professional literature prior to the mid-1970's. Before that time, many of the professions were male dominated. Thus, the pronouns "he" and "him" were used when referring to a professional. While gender-biased writing has decreased over the last decade, it still occurs in some material.

Every effort should be made to eliminate gender-bias in communication, even in professions that are still male dominated. Written documents have a long shelf life and, if they are well written, timely articles, they may be read long after gender equality has been achieved. Additionally, for those reading the material, gender-neutral writing will not suggest the exclusion of one gender from the profession.

A common example of biased writing is the use of the singular pronoun "he" to refer to a member of a profession. This may suggest that women are excluded or that women who are currently in the profession are not equal. Even for male-dominated professions such as engineering, gender–neutral wording should be used.

There are several ways of handling this problem. First, the sentence can be written to include both the masculine and feminine singular pronouns; specifically, "he and she" can replace "he," or "he" can be replaced by "s/he." Second, the sentence can be reworded to use plural pronouns that are gender neutral. Third, the sentence can be written with gender–neutral wording. A few examples best demonstrate these approaches.

Biased	*Unbiased*
After completing a design, an engineer should have his plans reviewed by another registered engineer.	After completing a design, an engineer should have his or her plans reviewed by another registered engineer.
An engineer should perform design work only in areas where he has the necessary education and experience.	Engineers should perform design work only in areas where they have the necessary and experience.
An engineer should accurately report his experience when soliciting work.	The experience of an engineer should be accurately reported when work is solicited.

To remove the bias in the first example, the feminine personal pronoun was included with the masculine form. In the second example, the bias was removed by rewording the sentence to the plural form. In the third example, gender was removed from the sentence. These examples show that gender bias can be eliminated from communications without loss of accuracy or meaning.

5.4 LIBEL

In nonlegal terminology, libel consists of a communication that dishonors a person's character. While this is not a common problem in professional communication, it should be avoided since, among other things, it may tarnish the profession. As such, it would violate professional codes of ethics. A criticism of the work of a professional may be judged as libelous if it damages the individual's professional reputation. This does not imply that the work of another cannot be criticized; however, criticism should be properly supported and professionally worded.

5.5 EXERCISES

1. Identify three human values and provide an example from engineering practice where each of the *values arises in an ethical conflict*.
2. Obtain a dictionary definition of plagiarism. What is the origin of the word? Discuss the similarities between plagiarism and kidnapping.
3. Discuss plagiarism from a value-conflict standpoint. Discuss how a person justifies plagiarism through rationalization. Illustrate using an example.
4. Assume that you are president of an engineering company. Compose a memorandum to your employees that expresses your views on plagiarism as it relates to professional activities of employees.
5. Obtain a copy of a copyright approval form from a publisher or editor of a journal. Evaluate its content.
6. Discuss why copyright approval is an ethical issue in a professional practice.
7. Identify the human values that arise in a gender-bias conflict. Discuss the balancing of the values.
8. Obtain a copy of published material of about one paragraph in length where gender bias is present. Rewrite the paragraph so that it is gender neutral.
9. Compose a short oral presentation that you, as president of your engineering firm, could give to your project managers on your views on gender bias.
10. Rewrite the following memo to eliminate sexist references:

TO: John Smith, Bill Cooper, Miss Mary Downs, Mrs. Ellen Anderson

FROM: Paul Allen

RE: Word Processing Pool

DATE: August 6, 1990

Yesterday I noticed the gals down in word processing were doing a lot of idle chatting, filing fingernails, etc. As you know, manpower requests are being cut severely, and any evidence of excess manpower is to be avoided. Please keep these little ladies busy! I'm counting on all of you guys to keep the work flowing.

11. Obtain dictionary and legal definitions of libel. Discuss the two definitions.
12. Provide reasons why libel is unethical as well as illegal.

CHAPTER 6 / **TECHNICAL REPORTS**

DO

- use the progressive outlining approach
- limit the introduction to a statement of the problem and the objectives
- make headings descriptive in formal reports
- recognize differences between formal and informal reports: goals, audience, time frame, format and scope
- write descriptive, yet creative, titles

DON'T

- worry about wording or rules of grammar when outlining
- use outline form for conclusions in a formal report
- use gender-biased wording
- ignore the importance of an abstract

6.1 INTRODUCTION

Convention usually sets the structure of a report. Companies have standard formats for reports, and academic institutions have established guidelines for theses and dissertations. The U.S. government has practically developed its own language and acceptable formats for written communications. Because there are so many different formats for reports, it is difficult to write a chapter on the structure of a report. When a company or college has a standard format, it should always be followed. Where a standard format is not established, the format given herein provides a useful structure for presenting technical material. Obviously, the final structure will depend on the audience, the intent of the report, and the specific content.

6.2 OUTLINING AND ROUGH DRAFTS

Would you expect a chef to cook a seven-course dinner without recipes or a menu? The chef may be able to prepare all seven courses, but the chance of a complete success is greater using the recipes and the

menu. The recipes summarize ingredients and outline preparation pro-
cedures, while the menu enables the chef to schedule the various tasks
in a timely manner.

Would you expect a movie director to begin filming without a storyboard? Without a storyboard to follow, the filming would be largely trial
and error—shoot, evaluate, and reshoot. The filming would be inefficient
and overly expensive, the parts of the final film would not blend well,
and the director would probably be frustrated throughout filming. The
storyboard enables the director to plan the filming and produce a coherent
story.

Would you expect a contractor to build a house without first obtaining
the building plans from the architect? Of course not! The building plans
help the contractor to properly sequence the construction and ensure that
the house will meet the specifications of the buyer.

A writer is like a chef, a movie director, and a contractor. The writer
needs a guide to plan the work and maintain focus on the end product.
An outline serves this purpose. It is to the writer as the recipes are to the
chef, the storyboard is to the director, and the building plans are to the
contractor. The outline represents the writer's opportunity to organize his
or her thoughts, to identify the objectives of the writing, and to summarize
the major issues to be discussed. It can also serve as a motivator. Without
the outline, the writer will probably work inefficiently, suffering through
many rewrites and ending up with an unsatisfactory product.

6.2.1 Progressive Outlining

Somewhere in our early education, all of us were taught how to
outline, and we probably have bad memories of the experience. While
these experiences may serve as a block toward outlining, experienced
writers know that outlining is the best way to get started. While learning
and practicing the outlining approach may be somewhat disconcerting
because of past negative experiences, mastering the progressive outlining
method will greatly improve your communication efficiency and enable
you to approach future writing exercises with a more positive attitude.

To begin, it is best to start with a small sheet of paper, such as the
back of a used envelope. The small working space will force you to focus
on the important elements and will prevent you from trying to use a
sentence-structured approach. The initial phase of outlining is the one
time in writing when the rules of proper grammar can be pushed aside.
The objective of the initial outline is to identify the important points that
you wish to communicate, not to produce a grammatically correct document. The first outline might be a series of phrases that identify: (1) the
problem; (2) the specific goal that underlies the work; (3) the approach
used in the work; (4) one or two important results; and (5) a major
conclusion. Fig. 9 shows an example of a first-pass outline for writing a

Problem:	Design method needed for controlling channel erosion below small urban reservoirs
Objective:	Develop design method
Method:	Computer model of system
	Develop design charts as a function of important inputs
Result:	Simple-to-use design charts
Conclusion:	Erosion control requires larger reservoirs than peak discharge control

FIG. 9.—*Initial Outline on Development of Reservoir Design Method for Control of River Erosion*

paper on developing of a method for designing reservoirs to control channel erosion in urban areas. It covers the important elements of the work from the beginning to the end. The brevity of a first-pass outline focuses the writer's attention on the important elements.

Immediately after completing the first outline, you can expand it into a more detailed second outline, which will provide slightly more detail and begin to provide some structure for a rough draft. As with the first outline, you should use brief phrases in the second outline. Fig. 10 provides an example of a second outline for the same problem used with Fig. 9. The second outline may begin to show a substructure that will evolve into subsections in subsequent outlines and drafts. This is illustrated in the first section of the outline of Fig. 10.

Subsequent outlines will provide progressively more detail and greater structure, but still use a hell-with-the-grammar philosophy. The expansion of the outline of the introduction can give greater detail about the problem (e.g., channel erosion is damaging to aquatic life and transports pollutants) and greater specificity of the objectives (e.g., to develop a design method based on hydrologic parameters and soil characteristics). Expansion of the literature review section of the outline can cite specific authors, their methodologies, and the input requirements for their models. The "approach" section may be separated into two parts, for example one on downstream erosion modeling and one on the reservoir routing. The "design method" section may also be partitioned into two sections, one on the final design procedure and one on its application. Greater detail will be given in the "conclusions" section, with topics such as the following included: (1) the added cost of design for erosion control; (2) needed changes in drainage control policies; (3) retrofitting of existing reservoirs; and (4) recommendations for public safety.

The outlines of Figs. 9 and 10 are examples of informal outlines. They have little structure and are limited to one level. For long, detailed reports,

1. Introduction	
Problem:	Design methods based on discharge control do not control erosion
Objectives:	Identify important inputs Develop design method
2. Literature:	Summarize commonly used peak discharge-based design methods
3. Approach:	Design-storm approach; Goncharov equation Describe hydrologic & soil inputs
4. Design Method:	Discuss development of design graph Discuss limitations Case study application
5. Conclusions:	Peak discharge control underdesigns for erosion control Erosion control is more costly Design will require assessment for both erosion and discharge control

FIG. 10.—*Second Outline on Development of Reservoir Design Method for Control of River Erosion*

planning is very important and a formal outline, with greater structure and more levels may be needed. A formal outline includes a list of major topics, with subtopics listed below each major topic. It may include a numbering system to provide a more systematic appearance. Very often, the listed topics and subtopics become the headings and subheadings in the report.

Fig. 11 shows an example of a formal outline. The example uses four levels to detail ideas, although the fourth level is used infrequently. The levels are identified by Roman and Arabic numerals and upper and lower case letters (e.g., IB2b). Alternatively, a tiered numbering system could be used, for example.

> 1. Introduction
>   1.1 Definitions
>   1.1.1 Discharge control
>   1.1.2 Channel erosion
>   1.1.3 Design criteria

If either of these two numbering systems will be used in the final report, it may be more efficient to use that numbering system for the outline.

The progressive-outlining approach has several advantages. First, it is very efficient and will enable you to get a lot of ideas down without worrying about format. Second, it will help you focus on the important

I. INTRODUCTION
  A. Definitions
    1. Discharge control
    2. Channel erosion
    3. Design criteria
  B. Problem
    1. Describe discharge-control design methods
    2. Detail effect of method on erosion
      a. increased erosion rates
      b. sediment accumulation in downstream reservoirs
  C. Objective
    1. Identify inputs for discharge control
    2. Identify inputs for erosion control
    3. Develop design method
II. LITERATURE REVIEW
  A. Discharge-design methods
  B. Channel erosion estimation
  C. Stormwater management policy criteria
III. DEVELOPMENT OF DESIGN METHOD
  A. SCS design storm
  B. Goncharov erosion model
    1. Required input
      a. Discharge characteristics
      b. Sediment characteristics
    2. Accuracy of estimated erosion
  C. Hydrologic conditions for design
  D. Erosion simulation
  E. Reservoir routing
  F. Computer model of hydrology and erosion
  G. Design method
    1. Development of design graph
    2. Input requirements
    3. Accuracy of design estimates
IV. APPLICATION OF DESIGN METHOD
  A. Verification tests
  B. Example design problems
V. CONCLUSIONS AND RECOMMENDATIONS
  A. Peak discharge vs. erosion control
  B. Cost of erosion-control design
  C. Policy recommendations
  D. Additional research recommendations

FIG. 11.—*Example of Formal Outline*

points. Third, it is easy to rearrange or expand upon the items in the outline. Finally, many of the items in the final outline can be used as headings or topic sentences in the rough draft.

At some point in the progressive outlining process, it will be evident that additional outlining will not add anything. Then a rough draft can be developed. The individual items in the last outline can be used to the develop the subsections and paragraphs of the rough draft.

6.2.2 Rough Drafts

Just as getting started with the first outline may have caused some apprehension, getting started with the first rough draft may elicit similar feelings. It is usually better not to start at the beginning with the introduction. It is probably best to start with the section that appears to be the easiest to write. Very often, this is the literature review or the data analysis section. Writing the least likable parts can be put off until the easier parts have been completed. Starting with the easier parts will generate positive feelings about completing the entire rough draft.

In the initial draft, don't worry about smooth transitions between paragraphs and sections. The transitional sentences can be inserted in subsequent drafts. The following are some additional tips for making a rough draft:

1. Don't worry about specific wording, spelling, or punctuation—these can be cleaned up in subsequent drafts.
2. Write quickly with the purpose of getting ideas onto the paper—avoid stopping so that writing momentum is not lost.
3. When one idea is completed, begin writing on the next on a new piece of paper. Then the pages can be easily rearranged prior to starting the next draft.
4. If motivation to complete a section wanes, start work on another, so writing momentum is not lost.
5. Don't be critical of your writing. It directs your attention away from the task of getting words down on paper.
6. Don't stop to draft figures or tables while completing the rough draft—just indicate that a table or figure is needed at a certain point.
7. Don't be tied to your last outline. As you write, expect that new ideas will surface; major changes can lead to improvements.
8. If you still have difficulties getting started, reread Chapter 2 on procrastination.

6.2.3 Revision

Once a rough draft is completed, put it aside for a few days. The hiatus will help you to take a more independent view of your first draft. In a way, it will be similar to having a friend review your work. Delaying the work may seem inefficient but it will actually improve your overall efficiency. The time should be used to make sure that your word processor is in good working order or to arrange for someone to type your final draft. The time could also be used to work on your bibliography and double check your citations and sources.

You should not expect to go from a rough draft to a final draft in one step as some parts of a report may require more revision than others. Good writers recognize the need for multiple drafts. Often writers approach each revision with a different purpose. One revision might focus on the transition sentences. Another revision might focus on developing good topic sentences, while another could be devoted to making sure that the implications of the observations are addressed. At some point, a revision should ensure that the sentences are clearly written and that there are no misspelled words. The following list includes items that should be considered in revising a report:

1. Objective is clearly stated in the introduction.
2. Headings are numerous and descriptive.
3. All technical terms are defined.
4. All notation is defined.
5. Words are chosen properly.
6. Gender-neutral wording is used (see Chapter 5).
7. Sentences are structured properly.
8. Proper transitions are made between ideas.
9. Paragraphs are structured properly.
10. Analyses are unbiased.
11. All figures and tables are included.
12. Each figure and table can be independently understood.
13. All ideas of others are properly referenced.

When you are satisfied with the report, it is important to have someone else review it. Too often, the writer forgets that he or she is totally familiar with the work; thus details that are necessary for the intended reader may have been omitted. Just as it is helpful to have a friend listen to a rehearsal of a speech, a friend's review of a written report can lead to many positive changes.

6.3 STRUCTURE OF AN INFORMAL REPORT

By the end of the first year of college, most students will have written several informal reports. The laboratory report for a freshman chemistry or physics course is an example. You should not get the idea that informal reports such as those for laboratory courses are something that only students must endure. In fact, laboratory reports are preparation for the endless stream of informal reports that most entry-level professionals labor over. Also, writing informal reports is preparation for the more detailed and more professionally important formal reports written by the experienced professional. Just as the freshman laboratory report is preparation for writing an informal business report, the M.S. thesis and Ph.D. dissertation are preparation for writing the formal reports required of those with post-baccalaureate degrees.

The informal report is usually written for a specific audience: the lab instructor or the project manager. The reader will be primarily interested in the results, and less in background information, knowledge development, or the general implications of the analysis. Therefore, the format of an informal report can be de-emphasized, with greater emphasis placed on speed and on transmitting specific information to the intended reader. These requirements often suggest an outline format.

Whether written in outline or prose form, headings are key to the success of an informal report. An informal report might include the following headings: Objective; Limitations; Assumptions; Methodology; Data Analysis; Results and Interpretation; and Conclusions. The headings should be descriptive, yet concise, since they serve as an outline for the reader as well as the writer. A consistent format should be used to present the headings.

An informal report can be in either prose or outline form, as long as the format is appropriate for the subject. Ease of reading is one of the most important criteria for selecting a format.

Fig. 12 shows a very simple example of an informal report. It is a laboratory report for a freshman course in physics. The specific problem is an analysis of data to test whether sample specimens obey a physical law. Because of the simplicity of the exercise, it is not necessary to put the tabular data into a formal table or the graphical data into a formal figure. Instead, they are placed in the report as needed. The headings are aligned along the left-hand margin and the content is aligned to the right of the headings to enhance readability.

In actual practice, the laboratory report may contain greater detail. For example, if the objective is to introduce the students to the measuring equipment (i.e., the voltmeter and the ammeter), then the students may be required to include more detail in the methodology section. Also, characteristics of the wire conductor (e.g., its length and diameter) may be included. The depth of discussion depends on the intended audience and the objective of the analysis.

6.4 STRUCTURE OF A FORMAL REPORT

The informal report is not always an appropriate form of communication. For example, when an engineer completes a study for a client, the

Objective: Determine whether an unknown conductor obeys Ohm's Law

Limitation: Tests conducted over the range of voltage (V):
                  $0.1 \leq V \leq 0.6$ volts

Assumption: Measurement errors in both V and the current (i) are
                  small and can be ignored.

Methodology: The conductor is connected between two points that are main-
                  tained at a constant potential difference for the duration of each
                  measurement. The potential difference is then changed and
                  the current measured.

Data: A sample of six measurements were made:

V (volts)	i (amps)
0.1	0.16
0.2	0.32
0.3	0.48
0.4	0.64
0.5	0.80
0.6	0.96

Analysis: The resistance R (ohms) was computed for each pair of meas-
                  urements: R = V/i, with a constant value of 1.6 ohms. The
                  values of V and i were also plotted using a rectilinear axis
                  system.

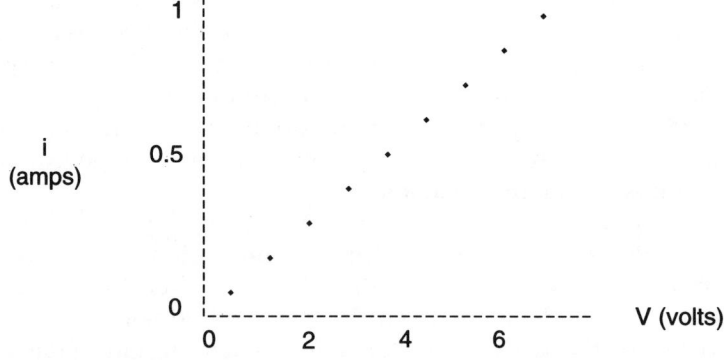

Results: (1) The resistance was constant.
                  (2) The plot showed a linear relationship between V and i.

Conclusion: Over the range of voltages studies, the conductor obeyed Ohms
                  law.

FIG. 12.— *Example of Informal Report*

complete details of the work must be described in a formal report. The outline form of Fig. 12 could not be used even though many of the topics shown in the figure would be addressed in the formal report. A more formal structure is called for.

A formal report (FR) differs from an informal report (IR) in many ways:

Goals
IR: Communicate results of a specific analysis.
FR: Communicate state-of-the-art knowledge and detailed results of a study.

Audience
IR: Often one person (lab instructor) or a small group (project team).
FR: An array of interested readers from the project manager to a client to outside professionals.

Time Frame
IR: A response to an immediate concern (i.e., short term).
FR: Long-term applicability.

Format
IR: Often a combination of outline and prose.
FR: Formal structure.

Scope
IR: Analysis of a specific problem.
FR: Detailed analysis of a general problem.

Probably the greatest difference between the two types of reports is their lengths. The informal report is usually relatively short; the formal report is much longer. For example, college theses and dissertations may be 500 to 1,000 pages. The length alone dictates using a formal structure for these forms of communication in order to maintain readability.

The informal and formal reports also differ in the amount of preparation involved. An informal report may require only one outline and a single rough draft, while a formal report may need several outlines and numerous drafts for each section.

Once the final rough draft is completed, it is necessary to put the report into final form. While many of the details will have been taken care of during the sequence of rough drafts, the final report should follow the established structure. The structure of a report will generally include elements such as the following:

1. Title page.
2. Abstract.
3. Table of Contents.

4. Text material of report.
5. References.
6. Appendices.
7. Bibliography (optional).

6.4.1 Title Page

The title page is used for documentation. Typically, it will provide, at the minimum, all bibliographic information including the name of the author(s), the title of the report, the organization responsible for distributing the report, and the date of publication. The title page may often include the name of the organization for which the report was written, the address of the author or distributor, and a brief abstract. Fig. 13 shows a typical title page for a company report. Fig. 14 shows the title page for a thesis. In addition to documentation, the title page adds an element of class to a report; it suggests to the reader that the author made the extra effort to provide an aesthetically pleasing appearance.

6.4.2 Abstract

An abstract is a vital part of any technical report; in fact, it may be the most important part. In professional practice, the abstract is sometimes called an executive summary. It is also referred to as a precis, synopsis, or brief.

Readers use the abstract to evaluate the relevance of the document.

STORMWATER MANAGEMENT:
A Comprehensive Study of the
Seneca Creek Watershed

Prepared for:
Montgomery County
Engineering Division

Prepared by:
XYZ and Associates, Inc.
3818 Southwest Parkway
Silver Spring, Maryland

Project MC038-3
August 1989

FIG. 13.—*Example of Title Page for Formal Report*

RISK-BASED CALIBRATION OF A HYDROLOGIC MODEL

by
Elizabeth W. Kistler

Thesis submitted to the Faculty of the Graduate School
of the University of Maryland in partial fulfillment
of the requirements for the degree of
Master of Science

1989

FIG. 14.—*Thesis Cover Sheet*

It is often the first part of the report read by anyone interested in the subject. It is, in effect, an advertisement for the report. A well written and complete abstract can sell the reader on the important information in the entire document. Conversely, the report may be rejected on the basis of a wordy, poorly written, abstract. Abstracts are also used as a substitute for reading the entire formal report.

A well-written abstract must identify the underlying problem (i.e., why such work was needed), the objectives of the work (i.e., what the work intended to accomplish), the scope (the parameters of the study), a brief statement of the results (i.e., how the work fulfilled a need), and a discussion of the implications of the results (i.e., how the work advances the state of the art or the value of the work to society). The abstract must be more than just a summary of the paper. A good abstract will entice the reader to read and subsequently use the material in the report. A poorly written abstract will suggest that the report is also poorly written, thus discouraging the reader from spending more time.

References should not be cited in an abstract because an abstract is meant to stand alone. Acronyms should also be avoided.

The abstract is usually the last part of a report to be written, but the first part to be read. The American Society of Civil Engineers (ASCE) *Author's Guide* states:

> In preparing an abstract for a paper, an author should follow these steps to obtain the desired coverage:
>
> 1. Review the summary and conclusions of the paper.
> 2. Review the text for additional information and examine captions of tables and figures.
> 3. Write the topical or opening sentence.
> 4. Complete the abstract by introducing the what, how, and why of the paper.
>
> The topical sentence should include the principal findings of the paper. It should not be a restatement of the title. Avoid starting with "this paper presents."

The ASCE Guide further states:

> Some general rules of style for writing abstracts follow:
>
> 1. Use direct statements rather than indirect ones whenever possible.
> 2. Use the present tense of verbs in describing conclusions and generalizations and in indicating the content of the abstract. Use the past tense only when describing the work done and observations made before the writing of the document being abstracted. The verb should follow the subject of the sentence as closely as possible.
> 3. Use complete sentences in all cases.
> 4. Avoid jargon and colloquialisms. Terminology accepted in a particular field, and commonly understood abbreviations, can be used. Phraseology that will confuse a foreign user should be avoided.

In technical writing, you will find three types of abstracts: (1) descriptive; (2) informational; and (3) evaluative.

The descriptive abstract is usually very brief, often 25 to 50 words. In a sense, it is just a list of key words in prose form. Because of its brevity, it is very appropriate for use with computerized abstract retrieval systems; thus, it often is totally independent of the formal report from which it was abstracted. The purpose of a descriptive abstract is primarily to let the reader know that a report containing work in a particular *speciality* is available.

The following is a descriptive abstract for a paper titled "Institutional Oversight to Enhance Integrity in Research:"

Because of the apparent increase in unprofessional conduct, an assessment of methods for reducing fraud in research is made. Causes of unprofessional conduct, the administrative responsibilities for research oversight, and the advantages of alternative forms of oversight are discussed.

This descriptive abstract, which totals about 40 words, indicates the problem (i.e., the increase in fraudulent research), the objective (i.e., an assessment of methods), and the scope of the paper (i.e., discuss causes, parameters, responsibilities, and forms). This descriptive abstract is composed of many key words (professional conduct, fraud, research, responsibility, oversight).

As an alternative to the above, judge the effectiveness of the following:

The importance of research to society is increasing. Oversight of research is needed to prevent fraud. Four forms of oversight are discussed: organizational codes of ethics; the appointment of an ombudsman for oversight; value education; and supervision and mentoring.

While this descriptive abstract uses approximately the same number of words and contains many of the same key words, it does not state the problem or the objective. The detail about the specific forms of oversight is probably unnecessary; it represents too much detail about a narrow part of the paper. The full scope of the paper is not presented because so many words are used up in the description of the forms of oversight. Therefore, the second descriptive abstract is not as good as the first one; it fails to inform the reader of the full scope of the paper.

The most commonly used abstract type is the informational. Its purpose is to provide facts about the formal report and its length can vary considerably. An executive summary may be 1,000 to 2,000 words and is intended for the executive who has no intention of reading the entire report. For brief formal reports the abstract may be limited to 150 to 200 words. However, the content is more important than the length. An abstract summarizes the principal ideas of the formal report; it does not include the subordinate, less important details.

Fig. 15 shows an informational abstract for a paper titled, "Institutional Oversight to Enhance Integrity in Research." The abstract begins with a statement of the problem (i.e., increase in unprofessional conduct) and the objective (i.e., evaluation of oversight of research). It then informs the reader that the issues of pressure and rationalization are discussed in the report. Other topics of discussion in the report are the pros and cons of institutional oversight, the responsibility and initiation of oversight, and forms of oversight. The abstract should conclude with a statement of the implications of the material discussed in the report.

Because of the apparent increase in unprofessional conduct, oversight of research is being considered as an alternative for enhancing professional integrity in research. Since those who have been involved in unprofessional conduct often cite pressure as the driving force underlying their conduct, two types of pressure, overt and perceived, are defined. A recent Institute of Medicine report has proposed a number of oversight mechanisms for medical research. Because similar oversight mechanisms may be given consideration in engineering research, the pros and cons of institutional oversight are addressed. Discussions on the responsibility for oversight and the initiation of oversight are provided. Four forms of oversight are discussed: Organizational codes of ethics; appointment of an ombudsman for oversight; value education; and supervision and mentoring. The voluntary adoption of internal oversight programs can reduce the likelihood of mandatory external oversight programs.

FIG. 15.—*Informational Abstract for: "Institutional Oversight to Enhance Integrity in Research"*

The third type of abstract is the evaluative. In addition to the type of information included in an informational abstract, an evaluative abstract includes the reader's opinion or evaluation of the quality of the work and the relevance of the results and conclusions. The length depends on the needs of the person(s) for whom the abstract was developed. For example, a junior member of a research team may have the responsibility to abstract current professional literature for the purpose of keeping the senior members informed about advances in the state of the art. Thus, detailed evaluative abstracts may be needed. As an alternative example, a junior county engineer may be responsible for evaluating existing policies and practice manuals on a particular problem. The senior county engineer may require only short evaluative abstracts so that he or she can inform the county council about recent advances in issues pertinent to their decision-making responsibilities.

An evaluative abstract is given in Fig. 16. The first part of the abstract states the objective, the scope, and specific conclusions provided in the formal report. The second paragraph provides an evaluation of the limitations of the report. If a company were interested in a water–quality parameter other than sediment, then the paper may have less relevance. If the importance of land–acquisition costs far exceeds those of construction, then the report may be of limited use. While the example shown in Fig. 16 does not address the perceived quality of the paper, evaluative abstracts can include the writer's opinion.

A framework for combining economic factors and the hydrology of detention basins is provided. The general development of an economic production function for water quality (sediment) and flood control is discussed. Example production functions are generated to compare water quality (sediment control only) and flood control. For the given example, the design of detention basins for downstream sediment is economically unwarranted. When compared to on-site detention facilities, regional detention structures appear to be more practical from an economic standpoint for water–quality control. Since sediment was the only water–quality parameter assessed, it is entirely possible that the design of a detention basin for water–quality control would be justified if the effects of all pollutants of concern could be quantified. Policy aspects of detention facilities that relate to the economics of water–quality control are also discussed.

Sediment is the only water–quality parameter evaluated. The case study uses a number of site-specific assumptions. The cost function is limited to construction costs, which ignores other costs such as those for land acquisition, planning, and design.

FIG. 16.—*Evaluative Abstract for "An Economic Framework for Assessing the Effect of Detention Basins on Sediment Control"*

6.4.3 Table of Contents

A table of contents, which must be a part of any lengthy report, should list the subdivisions and their page numbers in the report. All first-, second-, and third-order headings usually appear in the contents, although some abbreviated tables of contents use only first-order headings. If the headings are preceded by a number, such as those in this book, the numbers can be included in the table of contents. This book illustrates one format for presenting a table of contents.

A table of contents is more that just a means of finding specific information. It provides an overview of the structure and scope of the report, and is especially useful in reports without an index. The headings should be descriptive to help readers locate particular topics.

6.4.4 Text

The report outline serves as the initial form of the report structure. The outline headings correspond to the report chapters or sections. For example, theses text is divided into numbered chapters. For less involved reports, chapter or section numbers are omitted, but headings and subheadings are used to identify the different topics. For longer reports and theses, headings and subheadings should be used within every chapter.

They provide organization for the writer and enable the reader to understand how the sections fit together. Can you imagine trying to read this book if the headings and subheadings were omitted?

Headings and subheadings are a very important part of a report. Most importantly, they must be descriptive. The following examples show inadequate headings and descriptive alternatives:

Not descriptive	Descriptive
Model Development	Formulation of Transportation Planning Model
Data Analysis	Statistical Analysis of Maryland Water Quality Data
Results	Design Curves for Estimating Bridge Scour
Model Application	Application of CREAMS Model to Agricultural Areas
Recommendations	Policy and Design Recommendations
Discussion	Implications of the Model Study to Engineering Practice

In each of these comparisons, the descriptive heading is longer than its counterpart. While excessively long headings should be avoided, it is reasonable to have headings that are 50 to 60 characters in length. Nondescript headings such as those on the left in the above examples may be appropriate for chapter titles because the chapter includes a diverse discussion of material.

The format of the headings is also very important. Since the headings and subheadings help guide the reader through the report, the format should be clear and consistent. In some cases in this book, the headings are preceded by a number. A numbering system improves the readability of a report, but it also gives the report the flavor of an outline, which may detract from its formal appearance. Fig. 17 shows one possible heading format that is not based on a numbering system.

Headings and subheadings reflect the nature of the investigation being reported. The number of headings depend on the length, breadth, and complexity of the report. Figs. 18, 19, and 20 provide three examples of the major sections of reports for different types of problems. Fig. 18 shows the chapter headings for an analysis problem, such as those found in a thesis or dissertation; the report gives the details of a laboratory study and the subsequent analysis of the laboratory data. Fig. 19 lists the section headings of a synthesis problem. This might be applicable for either a thesis in which a model is developed or a report for a state agency describing the selection of a model for a particular technical problem. Fig.

The arrangement of titles and subdivisions within a chapter and their spacing is illustrated below:

First-order Heading

If there is but one rank of heading within a chapter, subdivisions should be indicated by an underlined centered heading in initial capitals, with two spaces between it and the last line of text above ("two spaces between" means that the typist triple-spaces) and one space between it and the text following (i.e., the typist double-spaces).

Second-order Heading

If there are two ranks of headings, subdivisions within the main rank are indicated by an underlined heading with one space above and below, and in initial capitals. The heading begins at the left margin.

Third-order Heading. If there are three ranks of headings, a third-order heading should be indicated by a heading indented 5 spaces (the number of spaces indented for a regular paragraph), underlined, initial capitals, followed by a period, and with the paragraph beginning on the same line and two spaces after the period.

FIG. 17.—*Headings for Formal Reports*

Introduction
Review of Literature
Instrumentation and Equipment
Description of Experiment
Presentation and Analysis of Data
Results
Summary and Conclusions

FIG. 18.—*Report Structure for Analysis Problem*

20 shows the section headings for a report intended to be a field guide for a planning problem; this would be comparable to a users' manual. For all of these examples, we would expect there to be subsections. Although the figures show generic headings, in practice the headings should be more specific and more descriptive.

A writer is certainly not required to follow the outlines of Figs. 18–20 nor will every report develop each of the items listed. However, actual

Introduction
Literature Review
Detailed Comparison of Existing Models
Model Development
Case Study Application of Model
Conclusions
Recommendations for Implementation

FIG. 19.—*Report Structure for Synthesis Problem*

Problem
Alternative Planning Practices
Recommended Practice
Example Applications

FIG. 20.—*Report Structure for Application Study*

headings should be very descriptive of the section's content. Dividing the text into the following sections helps the author organize the report and helps the reader follow the path taken in conducting the work.

Introduction. The introduction should include a concise statement defining the problem, a brief history leading to the problem, and the purpose of the work. The introduction should not be a review or summary, and it must not contain conclusions of the report. Instead, the introduction should set the stage, define limits, and describe why the work is needed. The underlying purpose of an introduction is to entice the reader to continue reading, not to give results that would discourage the reader from continuing.

Review of the Literature. If applicable, this section should indicate briefly what has already been reported in the literature concerning the problem, the difficulties that may have been encountered by other researchers, criticisms of previous approaches, etc. If the literature is not extensive, the writer may wish to treat this section as a separate part of the introduction section. The review of the literature may be presented as a temporal summary of the progress of the state of the art. If the literature base is extensive, the literature review section will be divided into subsections.

It is very important to distinguish between your work and efforts by others. The literature review section provides you with the opportunity to report on the work of others that has set the stage for your work. By placing the work of others in the literature review, your current work can be put in the other sections of the report so that the reader will clearly distinguish your work from that of others.

Instrumentation and Equipment. This section is common when laboratory work or a field study is involved. It should be described and illustrated in sufficient detail so that a skilled person could set up the apparatus and duplicate the work. Reports in which laboratory work was not necessary may omit this section. If the research involves extensive computer work rather than a laboratory problem, a section may be devoted to the computer facilities, set up, programming requirements, software, etc.

Description of Experiment. This section should review the physical laws that underlie the experiment, the steps taken in conducting the experiment, the variables (or inputs) involved and how values for each were decided upon, and the expected outputs (but not the results). Sufficient detail should be given so that the experiment could be reproduced by others. Any assumptions made should be clearly stated.

Presentation and Analysis of Data. The experimental design usually involves measuring outputs. The measured values, which may include measurement error as well as sampling variation, must be summarized and analyzed. For example, mean values and ranges may be computed (see Chapter 4). Basic statistical tests may be made on the measured data. These analyses should be summarized in sufficient detail such that the reader will understand what data were collected, what analyses were made and why, and the implications of these data summaries and analyses. Extensive tables of data can be placed in appendices to the report.

Results. This section consists of a description and interpretation of data obtained in the study. Tables, charts, and curves of a *summary* nature should be included in this section, as well as photographs or sketches that add to the presentation. Most research data, however, should be reserved for an appendix to which readers are referred if they wish to verify statements. The section of results should point out highlights and items of significance in the data *and allow the author to focus attention on the most important findings.* The "results" are the hard facts presented in the actual data. *This section should also include the writer's interpretation of those facts.*

Conclusions. This section should summarize the results of the research presented in the report. The results should be based on factual findings. Each separate conclusion should be discussed in a logically sequential order. Because conclusions are often lifted out of the context of the report and quoted without the explanatory material, care should be taken to compose each conclusion to ensure that it does not imply a broader scope than is intended and that it includes the necessary qualifications. *The tendency to present the conclusions in outline form should be avoided* because it is important to state the implications of the findings. Thus, one-sentence statements summarizing the important findings may be inadequate to portray the scope of the investigation.

Appendices.   The purpose of an appendix is to present those details of data that will verify the summary statements reported in the text but that would obscure the development of the presentation if included in the main body of the report. The appendix or appendices should contain the bulk of the research data or findings as embodied in the tables, diagrams, sketches, curves, and photographs. They should contain such items as sample computations and derivations, computer programs and output, and material that is too voluminous for inclusion in the main report. A "Glossary of Abbreviations" may be included as an appendix. Also, an appendix entitled "Notation" may be used to list the definition of all mathematical symbols used in the report.

Each appendix should be indicated by a letter (e.g., Appendix A), include a title, and have a cover page.

6.5   FEASIBILITY REPORT

Suppose two proposals have been put forth for increased airport facilities in a major urban center. One possibility is to enlarge the existing airport; another is to revamp a nearby Air Force base scheduled for closing. Each proposal would involve major expenditure; therefore, choosing the wrong proposal could be very costly.

A feasibility report is a document produced by an employee or group of employees who gathers data on alternative solutions to a problem, establishes criteria for judging the solutions, and recommends a plan of action. A feasibility report gives decision-making information to management that allows them to make successful and cost-effective decisions. A feasibility report is an excellent example of what Nesbitt means when he says, in *Megatrends*, "Information is today's valuable commodity."

There are many situations in which an employee might be required to write a feasibility report. An organization may plan to purchase new computer software or hardware, relocate its manufacturing activities, or develop a new product. In industry, change is inevitable, and feasibility reports allow an organization to make cost-effective and efficient changes based on a rational analysis of all alternatives.

Each organization will have its preferred way of writing feasibility reports. However, most reports contain the following sections: Introduction, Purpose, Definition of Problem, Criteria for Selection, Alternative Solutions, and Recommendations.

Introduction.   The introduction sets the tone for the report. It should be clear and well organized and should prepare the reader for exactly what will be discussed in the body of the report.

Purpose.   The purpose for writing the feasibility report should be clearly stated. The various alternatives should be briefly mentioned, and the writer's intention to recommend a solution should be stated.

Definition of Problem.   Some background or historical information should be discussed in this section. Circumstances or situations that led to the need for change should be discussed. It is important to convince the reader that a problem exists before solutions can be proposed or implemented.

Criteria for Selection.   In order to judge all alternatives fairly, the basis for judgment (the criteria) must be carefully developed. Cost is generally the first criteria to be considered. All types of costs—initial, operating, and maintenance—should be included in estimates. Capability is a second important critieria. What must the new project do or achieve? Returning to our example of the airport, we might ask the following questions:

1. How much would a newly constructed airport cost?
2. How much would renovations to an existing facility cost?
3. How much air traffic could an expanded site handle?
4. How much air traffic could the renovated military facility accept?

Asking and answering these questions allow us to develop cost and capability criteria for the feasibility report.

Alternative Solutions.   Each solution should be carefully described. A balanced presentation should be given to each option. Biased presentations can lead to non–optimum decisions. Each solution should address the cost and capability criteria.

Recommendations.   A plan of action should be recommended from a careful analysis of the problem, the criteria for decision-making, and the alternative solutions.

Audience.   As in all report writing, a feasibility report must address its specific readers. The technical background of the reader must be taken into consideration when illustrations and graphics are planned. The level of technical information should be also be appropriate for the reader.

6.6 WORD PROCESSING

Throughout one's undergraduate program, occasional papers or reports are required. Very often, these reports are informal, handwritten, and intended only as a summary of work performed. In graduate school, students spend a much greater amount of time writing reports, and these reports represent a greater portion of the grade. Therefore, their appearance is given greater weight. These papers and reports are expected to be professional in appearance. The currently available technology makes this task more manageable than in the past.

Two options are available, the personal word processor and the word–processing software packages that can be used on personal computers. The former is less expensive, while the latter is more flexible and powerful.

Few books on technical writing discuss the use of word processors. Even though an abundance of word–processing manuals are commercially available, they are usually geared toward a particular software package. Books on technical communication often omit discussions of word processing because of the rapid pace of technological advancement in the industry.

This chapter will not cover the specifics of word processing. It is meant as a guide to selecting a word processor, creating a first draft on the word processor, and making revisions to drafts of the paper or report.

6.6.1 Word–Processor Selection

In selecting word–processing software, there are a number of important criteria to use. You will need to consider the different uses you will have for the software, compatibility with software at work or school, and availability or ease of use of graphics.

There are many word–processor software packages on the market. If you intend to do your writing at the workplace or at school, you will have to learn to use the available word processor. If you work at home, you may want to use the same software that is used at work so that you can print your document there or have a secretary make revisions without transferring the document to a different system. The transfer can result in a loss of information and be very inefficient.

If you are not concerned with compatibility with the word processors at work or school, there are other options to consider. If you plan to use mathematical or engineering symbols, your choices of appropriate software are more limited. Some processors have a wider variety of symbols so they handle mathematical and engineering symbols better than others. Greek symbols are often used in physics and engineering. You will not want to type these in with a manual or electric typewriter after you have printed out your document stored in the word processor; therefore, make sure that the commonly used symbols exist in any word processor software that you purchase.

You may also need to combine text and graphics into a report. There are several ways to do this. If you are working on an IBM–compatible machine, you will likely use separate software to create the graphics, then combine the report and graphics after each has been printed. If you are using an Apple Macintosh computer, combining text and graphics is a bit easier. Both can have a very professional appearance.

6.6.2 Composing on the Word Processor

When it's time to write a paper, you have two choices; you can use a pen and paper or you can use a keyboard and word processor. Both forms of writing require the same key ingredient: an outline. You should never sit down to write at a keyboard without first making an outline, tempting as it may be. No matter how you move those sentences and paragraphs around, they will not flow well if you started without considering the big picture.

Keyboard composition can be either a help or a hinderance. If you are a relatively good typist, keyboard composition is a help. You will have everything typed without the need to write the paper manually and then type it or have it typed. If you can't type well, keyboard composition will be a hinderance, forcing you to think more about typing than writing. When you first draft a paper, you should not worry about sentence structure, spelling, typing, or other mechanical issues. You want your creative juices to flow freely. If you have to think about typing, this will not occur.

Two of the great benefits of word processors are the spell-checking and thesaurus features. Spell-checking is terrific. You'll never again have to proofread a paper for spelling. Right? Wrong!! Sometimes a word that is actually misspelled will be passed over by the spell-check. For example, if you mean to type the word "by," but instead type "be," the spell-check will not pick up the misspelled word because "be" is a word. So do not fail to allow time to proofread your work prior to submitting the final copy.

The built-in thesaurus offers the writer alternative word choices at the touch of a button. Varying the words in a paper can make the writing more interesting to read; however, overuse of a thesaurus is usually obvious. Rather than using the thesaurus frequently, it is better to use it only occasionally and try to restructure the sentence at other times.

6.6.3 Making Revisions

Making revisions on a word processor is, of course, a simple matter. You may make additions, delete sections, or move sentences and paragraphs around. However, it may be best to print the draft of the paper out first, read it carefully, and make notes on the paper before making revisions. It is sometimes difficult to sense the flow of a composition on a computer screen.

6.7 EXERCISES

1. Choose a partner in class. In the next class session, take notes which will enable you to write an outline of the lecture. Exchange outlines with your partner and annotate each other's outlines. Discuss the strong and weak points of each outline.

2. Find an editorial in a professional trade publication of your technical speciality. Develop a detailed outline of the topic that could be used as the start for a more detailed discussion of the topic.

3. Using a laboratory report that you wrote for a course, develop a detailed outline for making an oral presentation of the laboratory work.

4. Develop a detailed outline of a report that explains to a group of high school students one of the following topics: friction, thermal conductivity, nuclear energy, the behavior of molecules, static electricity.

5. Reread one of your previous writing assignments three times. On the first reading, focus on the first sentence of each paragraph, making certain that it introduces what will be discussed in that paragraph. During the second reading, make sure that each sentence in each paragraph is in a logical order and that it is pertinent to the topic sentence. On the third reading, add descriptive headings where they are necessary or change existing headings that are not descriptive.

6. Obtain several papers from professional journals in your field of interest. For each paper list the headings and subheadings as they appear. Do the headings clearly identify what will be discussed? For headings that are not descriptive, replace them with more descriptive headings.

7. Write an informal report modeled on the report in Fig. 11. Choose a danger or a health problem in your school or community. Include your objective, assumption, methodology, data, analysis, results, and conclusion.

8. Develop a laboratory report using the form of Fig. 12 that summarizes a laboratory experiment to estimate the coefficient of linear expansion of a steel bar.

9. Go to the reports section of your university library. Read and outline a formal report that has been published in an area in which you are interested.

10. Write both a descriptive informational abstract (about 150 words) for a previous writing assignment, a chapter of a book, or a published paper that does not have an abstract.

11. Using a professional journal that includes abstracts of papers, find two abstracts, one that is well written and one that is poorly

written. Explain why the one is well written and the other is poorly written. Rewrite the poorly written abstract.

12. Obtain three reviews of word–processor software packages from printed material, such as newspapers and software-oriented journals. Develop criteria that can be used to compare the software.

13. Find an old term paper that you wrote on a typewriter. Study the format and make notes on ways in which the word processor could have produced a more effective format.

14. Find promotional information on two similar word–processing software packages. Write two to four paragraphs, comparing (similarities) and contrasting (differences) the two packages. Assume that you are writing this for a manager who is computer literate and uses older software for small communication tasks (e.g., memos, minutes of meetings). Then rewrite the paragraphs for a manager who is computer illiterate. How does the audience influence the content and form of the communication?

15. Investigate two home entertainment options, a C-band satellite dish and cable television installation. Write a feasibility report and make a recommendation.

CHAPTER 7 / ORAL COMMUNICATION

DO

- prepare as far in advance as possible
- rehearse in front of friends
- make optimum use of visual aids
- entertain your audience while you are educating them
- make eye contact with your audience
- vary the speed and loudness of your voice

DON'T

- read your speech
- use filler words: uh, uhm, you know
- limit your attention to just one part of your audience
- stand in one place—movement provides variety and helps maintain the audience's attention
- exceed the allotted time

7.1 INTRODUCTION

How much of a young professional's time requires oral communication skills? For many, it is as high as 30 percent. As one rises in the managerial ranks, the percentage will increase. Furthermore, the rate at which a young professional advances up the corporate ladder is highly correlated with the individual's ability to communicate.

The need to communicate orally is not usually evident from the college curriculum. "Public speaking" courses are rarely required in professional-school curricula. So students believe that the emphasis in professional life will mimic the emphasis in their college courses, which, for the most part, are oriented toward technical skills. Students are often surprised and unprepared when they graduate to find that they lack the oral skills required in professional life.

Probably the first instance requiring these skills is during the job interview (see Chapter 11). While the letter and resume sent before the interview are important, the deciding factors will probably be the knowledge, confidence, and maturity exhibited during the meeting—all of which are orally communicated. The individual who interviews well always has a definite advantage.

In professional practice, there are innumerable situations in which the young professional will need good oral communication skills. Initially, he or she will have to provide oral summaries of technical work to managers and other professionals working on other aspects of the job. These may be informal briefings or regularly scheduled meetings. In time, meetings with prospective clients to win jobs will be necessary, as well as presentations to existing clients to report the status of work. Presentations to the public may also be required, and presentations at professional society meetings are sometimes a way of promoting your ideas and the company's work. Success in professional life will be greatly influenced by the speed at which you acquire good oral communication skills.

There are many types of oral presentations. One is an impromptu summary of progress on a specific work assignment given to a small group sitting around a conference table. Such an impromptu speech can be very important to your professional advancement. In some cases, a presentation will be more formal, made standing before a group and allow the use of the blackboard or a flip chart. In some cases, you may only be given an hour's notice before such a presentation.

Presentations to larger audiences are common in professional life. Often, engineers are required to make a presentation to the county council on the impact of a project being developed by the company. The audience might be a homeowners' group and extensive questioning might follow by the audience or council members. For such a presentation, there will usually be sufficient lead time to prepare a formal presentation. Advancement in your professional life may also require making speeches before large audiences at seminars or conventions; here, audience participation and questions may be minimal. This chapter focuses on presentations where preparation is possible. However, many of the principles apply even to impromptu talks.

This chapter provides information on the four steps in the oral communication process: (1) Formulating the presentation; (2) compiling the material for the presentation; (3) rehearsing; and (4) making the presentation. It is important to recognize that a poor presentation almost always results from failure during the first three steps, not the fourth step. If sufficient attention is given to the preparatory steps, then the actual presentation will most likely be successful. Proper attention to the first three steps can also help reduce nervousness, which is usually the number one concern of the novice.

7.2   FORMULATING THE PRESENTATION

The best way to formulate a presentation is to answer the question, "What major point(s) should be made?" By focusing on the major conclusions of the presentation, one can then prepare to educate the audience.

Knowledge of the audience is essential to a successful presentation. The first step in formulating a presentation is to assess the knowledge level of the audience. It is also necessary to know the range of their knowledge. It is easier to prepare a presentation for a homogeneous group, one in which most of the people within the group have a similar knowledge level. A class of college freshmen is an example of a homogeneous group. For a heterogeneous audience a presentation formulated for a high knowledge-level group will be ineffective in communicating your work to those who are not totally familiar with the general topic on which you are speaking. Conversely, a presentation formulated to reach those in the audience who have limited knowledge of the subject matter may bore those with a greater understanding of the topic. In general, when faced with a heterogeneous audience, it is probably best to try to include material that will appeal to and educate as many knowledge levels as possible. This is not easy, but it is necessary for a generally successful presentation.

The ultimate objective is to educate the audience about the results of your work. Having identified your major conclusions and acknowledged the background of the audience, you list the specific objectives. It is usually most efficient to make a short outline-form list of your objectives which very often you can use as part of your introduction. It may also be useful when preparing your visual aids.

For most presentations, your objective will be to educate the audience. You will be informing them either of your progress or the results and conclusions of your work. You have been selected to speak because you have knowledge that others need.

Alternatively, the objective may be to persuade the audience. For example, the presentation may focus on the qualifications of your company, the intent being to persuade a client to contract your company for work. Or, you may serve as an expert witness in a legal case and your objective will be to persuade the jury or judge that your technical assessment is correct. For both objectives—education and persuasion—preparation and a complete understanding of the material are essential.

Most oral presentations have a time constraint which limits the amount of material presented. It is important to acknowledge this time constraint when formulating a presentation. Otherwise you may attempt to cover too much material. An audience can grasp only so many ideas in a given time period and their comprehension will drop off considerably when confronted by too many new ideas. The underlying lesson is to present only the most important ideas and to use these ideas as the theme of your speech. The audience will retain the primary points if they are emphasized through repetition and not interspersed with less important ideas.

7.3 DEVELOPING THE PRESENTATION

A very efficient way of developing a presentation is to use the progressive-outline approach. With this method, a very simple outline of four to six lines is made to address the following questions:

1. Why was the work done? (State problem and goal).
2. How was the work done? (State solution method).
3. What findings resulted from the work? (State one or two major conclusions).
4. What do the results imply? (State the implications of the work).

The second outline should be about twice the length of the first outline, again concentrating on the major points. Successive outlines continue to double in length, with progressively more detail of the study included. This progressive approach of outlining ensures that you will focus on the important ideas and that the topics are in the proper sequence. The progressive outlining approach, including examples, is discussed in greater detail in Chapter 6.

In progressing through the various stages of outlining, consideration should be given to the use of visual aids. Transparencies, videotapes, computer demonstrations, and 35-mm slides vastly improve audience comprehension; they also reduce speaker nervousness because they divert the audience's attention away from the speaker toward the screen. Visual aids can also serve as cues for the speaker, thus eliminating the need to use notes. The topic of developing visual aids is discussed in Chapter 8, but it is at the presentation-development stage when ideas on these aids should first be considered.

When developing the outline, give special emphasis to the introduction and conclusions since these are the most important parts of the presentation. The introduction should concentrate on the problem (i.e., reason for doing the work) and the objectives; the introduction should not present the results of your work. During the conclusion of the presentation, you should summarize the major findings of the work, show how the study met the objectives specified in the introduction, and provide a discussion of the implications of the results.

The first minute or two of a presentation is the most important part because it is during this short time span that audience attentiveness is at its peak. During this brief period of time, a significant proportion of the audience will be deciding, possibly without conscious awareness, on the extent to which they will concentrate on the remainder of the presentation. If the introduction is boring, many in the audience will allow their minds to wander to more gratifying thoughts, such as their own work or where they will lunch. For this reason, the intent of the introduction should be to capture the audience's attention. It should create enough interest to hold the audience throughout the speech.

How can a speaker capture the audience's attention? An educational introduction is necessary but not usually sufficient. You must entertain, not just educate. After-dinner speakers often use a story or a joke to capture the audience's attention. If used properly, such openings persuade those in the audience to continue to pay attention; thus, the introduction has fulfilled its basic purpose. But a joke or story is often not appropriate for professional presentations, especially technical presentations with a limiting time constraint. Fortunately, there are effective alternatives for introductions to technical presentations.

Recognizing the diversity in interests in a heterogeneous audience, a series of slides showing the physical facilities related to the presentation's subject can be a very effective introduction. For example, a presentation of a design practice to improve public safety may begin with some slides showing damage caused by past failures. Such slides can be quite captivating and can simultaneously educate the audience about the problem. Similarly, a presentation on the loss of wetlands or other environmentally sensitive lands could begin with slides showing a couple of before–and–after comparisons. Such introductions are effective because the slides will appeal to the wide range of interests and knowledge levels in a heterogeneous audience. The slides further suggest that the speaker has the maturity to put his or her work into the broader social perspective.

While the introduction to a presentation is the most important part, the conclusion ranks a close second. This is your final chance to make a lasting impression on the audience and to make your central point. Therefore, it is important to make sure that you have their undivided attention. One way of getting their attention is to say, "In conclusion, . . ." or "In summary, ..." For some reason, these magic words recapture the attention of whose minds have wandered. Another way is with a visual aid that includes the word "conclusions" in bold letters.

Having gotten their attention, you will want to hit them between the eyes with your major point. This may be a technical result of your work or a recommendation for action. If you are trying to convince the audience that your approach is the best alternative, you might want to summarize the advantages of your approach. You can describe the benefits of your approach over the existing approach to a problem. Whatever your major point, be brief. Lengthy conclusions are not effective.

Once you have developed a detailed outline of the speech you can plan your written version of the presentation. However, there are advantages in not writing out a detailed version of the oral presentation. First, this is very time consuming, and it may be better to use the time to prepare the visual aids and rehearse the speech. Second, when a written version of the presentation is available, there is a strong tendency to read the speech from the written copy. A speech should never be read (see Section 7.7).

7.4   REHEARSING THE PRESENTATION

Once you have written the outline for the speech and selected the visual aids, you should rehearse. Proper rehearsal will help identify weaknesses in the presentation, help to overcome pre-speech nervousness, and ensure familiarity with the flow of the presentation. If possible, the presentation should be rehearsed at the same location where it will be given; familiarity with the surroundings and the equipment will help reduce nervousness during the actual presentation. If you cannot rehearse at the location, rehearse at several different locations so you will be accustomed to presenting in unfamiliar surroundings. If possible, have a friend or colleague videotape your rehearsal of the presentation. Watching the videotape is an excellent method of self-evaluation.

Time is always an important constraint. When a specified length of time is allotted to a presentation, the speaker must not exceed that period. Failure to comply with the time limitation often means that there is insufficient time for the audience to ask questions or that time allotted for subsequent presentations is shortened. Always avoid this lack of professional courtesy. When rehearsing a speech, record the time so material can be deleted if necessary. Since your reading speed far exceeds your rate of speaking aloud, the presentation should always be rehearsed aloud, not just read in silence. A silent rehearsal takes far less time and you will underestimate the time the presentation will actually take.

7.5   MAKING THE PRESENTATION

The first point to remember in public speaking is that nervousness is to be expected. Nervousness, however, does not start the night before or five minutes before the presentation; its roots are in the development and rehearsal stages of the presentation. An improperly developed or inadequately rehearsed speech is good reason for being nervous. But there is hope!

First, except in extreme cases, nervousness will not be detected by the audience; rarely does someone visibly shake while making a speech. A speaker may tend to talk too fast, avoid eye contact, or get a dry throat, but these will not spell the disaster of a presentation.

Second, nervousness usually subsides as the speech progresses. It is usually most acute at the beginning of the speech, so the introduction should be given special attention during preparation and rehearsal.

Finally, some nervousness may actually be helpful if it makes the speaker concentrate more on what he or she is saying. By concentrating on every word, the speaker's mind is taken off both the audience and any pre-speech nervousness. This should also help avoid the tendency of a nervous speaker to talk too rapidly. So nervousness is reduced by concentrating, and it can be used to advantage.

Eye contact with the audience has already been briefly mentioned, but its importance warrants specific attention. Making eye contact with the audience helps to hold their attention by bringing them into the presentation. It also enhances the audience's confidence in the speaker. It is part of the body language of communication (see Chapter 12). Mastering the art of making eye contact with the audience is one of the key elements in developing self-confidence in making presentations.

If you have ever heard a speech given in a monotonic voice, you will know that even a well prepared speech can fail when poorly presented. Just as slides or other visual aids vary a presentation, so can modulating your voice. Raising or lowering the volume can heighten the attentiveness of the audience. Similarly, increasing or decreasing the speed can be an attention-getter. Significant changes in either volume or speed are a good way to emphasize a point. Similarly, moving out from behind the podium will get attention and give you the opportunity to make a point.

One other note about delivery. Avoid using fillers such as, "I mean," "you know," "uhm," or "uh." These distract the audience and can actually be annoying. Speaking slowly and concentrating on your words helps avoid the use of fillers.

In addition to voice control, body control can influence speech effectiveness. A stiff, inanimate stance bores an audience in the same way as does a monotonic voice. Crossed arms signal a defensive posture just as fidgeting with a pencil indicates nervousness. Body control is another part of kinesics, which is discussed in Chapter 12.

Finally, it is worthwhile remembering the motivation loop (see Fig. 21). Feeling engenders action, with the action engendering other feelings. This loop can serve either a positive or a negative purpose. If one has negative feelings about making a presentation, then the probability of negative action (i.e., a poor presentation) increases, and the negative action will engender negative feelings for future presentations. It can be a vicious cycle, but it can also be a positive cycle. If an oral presentation is approached with positive feelings and combined with proper preparation and rehearsal, then the probability of positive action (i.e., a good speech) increases. Positive feelings engender positive actions, which then engender positive feelings about making future oral presentations. The positive path of the motivation loop is certainly the easiest.

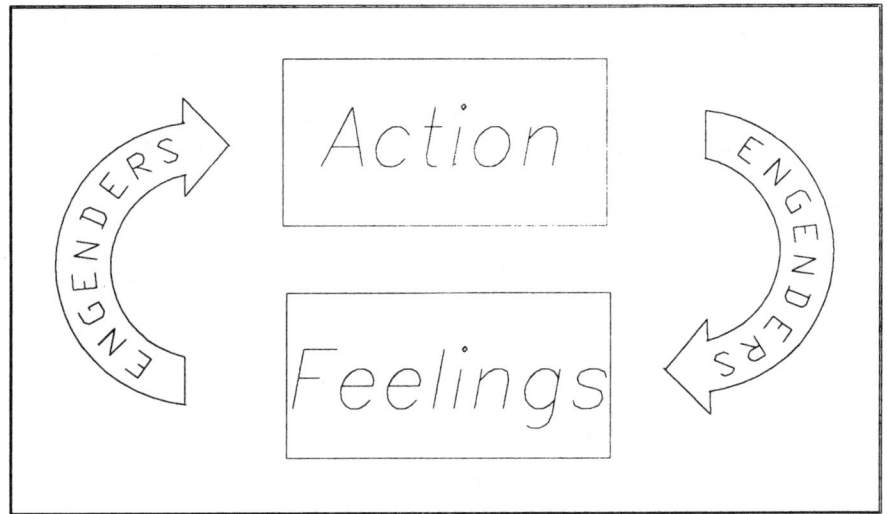

FIG. 21.—*The Motivation Loop*

7.6   RESPONDING TO QUESTIONS

Those with little experience making oral presentations have almost as much fear of the question-and-answer period as they do of the first few minutes of the presentation. The fear arises because they recognize that they will not have control of the questions and often believe that their lack of command of the technical subject will be exposed through in-depth questioning. Just as good planning and rehearsing will put a speaker in control of making an oral presentation, there are ways of gaining control of the Q-and-A period.

It is important to make sure that everyone in the audience has heard the question. If the questioner speaks with an accent or sound does not carry well in the room, then it is best to repeat the question or, if the question is lengthy, the speaker can paraphrase the question. Repeating the question has the added advantage of buying time to formulate a response. It is similar to re-reading a sentence or paragraph of a written report; it will help you understand the question better.

Once a question has been asked, the speaker has control only when he or she has control of his or her thoughts. If panic takes over, then he or she will not be able to respond effectively. If a multipart question is asked, the speaker should try to respond to each part separately, possibly responding to the easiest part first. If a very broad question is asked, the speaker should try to redirect the question into a more specific question that relates to the material covered in the speech. If a very uninformed question is asked, the speaker should not embarrass the person who asked it; instead, the speaker should rephrase the question into a more sensible question and give a brief response to the revised question. If a somewhat sarcastic remark is made, such as when the speaker and the questioner have used drastically different assumptions to solve a problem, the speaker can try to defuse the confrontational atmosphere by pointing to the similarities in the alternative approaches and to positive aspects of the method used by the speaker. In any of these cases, the speaker can get control of the situation by trying to identify the fundamental concept that underlies the question, that is, to simplify it. Don't be afraid to take a few seconds to think about the question; an immediate reply is not expected. Use the few seconds to develop an organized response.

7.7   READING A SPEECH

Why would someone want to read a speech? First, it reduces the time required to rehearse the speech; it would only be necessary to write out the speech and then review it to make sure that it flows smoothly. Second, it serves as a crutch for someone who expects to be very nervous when making the presentation; they expect their nervousness will cause them to forget what they are supposed to say and reading will eliminate this possibility. Third, while reading, the speaker does not have to look at the audience, which can be a significant cause of nervousness. Reading a presentation or using notes excessively causes a number or problems. First, the presenter frequently loses his or her place in the notes, causing confusion and detracting from the flow. Losing the place in the written copy causes panic since most of the preparation time was spent reading rather than learning the speech. Second, reading minimizes eye contact between the presenter and the audience, thus reducing the audience's feeling of involvement in the presentation. Third, reading suggests a lack of respect (i.e., the speaker has not taken the time to prepare properly) and that the speaker does not have a good grasp of the material.

In fact, it is easier to make a presentation without notes than with them. Visual aids, used properly, can serve as an outline and prompt the speaker's memory on the topics to be discussed. With good visual aids, even 3 × 5 note cards are not needed. It is important to address the issue of using notes or a written manuscript when discussing speech rehearsal since it is then that the speaker will be preparing both visual aids and any notes that must be used. Notes should be avoided, if possible. If necessary, they should be limited to keywords that prompt the speaker's memory.

7.8   EXERCISES

1. List the instances in which you were required to make an oral presentation during high school and college. Summarize the ex-

tent to which you prepared for these presentations and evaluate the success (or failing) of the presentations.

2. Using a written laboratory or project report, develop a 10 minute oral presentation of the work. After completing the presentation, provide a summary of the steps you used in developing the speech.

3. Select a subject, perhaps from a recent report, develop an outline to present the material to (1) a group of peers; and (2) a group of high school students; and (3) a group of interested professionals from various fields.

4. Working in a group of two or three, choose a topic for a brief oral report. Role-play a poor presentation and then present an excellent report.

5. Develop a creative introduction for a presentation aimed at explaining thermal conductivity to a nontechnical audience.

6. Write a critique of a speech or oral presentation on television. Identify strong and weak points in the presentation.

7. Observe how one of your teachers responds to questions. Discuss how the teacher could improve his or her technique.

8. For one week, keep a log of all lectures, presentations, or speeches that you attend. In the log, identify the strong and weak points of each speaker.

9. Practice an upcoming speech before a group of friends. Ask one friend to watch for your eye contact and other body movements. Ask another to listen for technical content. Ask a third friend to give special attention to visual aids and a fourth to check the overall flow of the talk. Have them provide a written assessment of your speech. Then make a written list of ways that you can improve your oral communication skills.

10. While practicing your presentation, practice once standing very still. Then practice moving and pointing to visual aids while speaking. The second practice should be more interesting and pleasant.

11. Discuss why it is improper to take longer than the allotted time to make an oral presentation.

12. Discuss the disadvantages of reading a speech.

13. Using your college course catalogue, identify courses that provide instruction in oral communication.

14. Prepare a speech to present a technical process or piece of equipment. As part of the preparation for the speech, answer these equations:
    a. What is the objective?
       Teach; persuade; stimulate
    b. What do you know about the audience?

How many people; what background or experience level; what knowledge of the subject

c. What do you know about the situation?
   Size of room and amount of seating; lighting and acoustics; time available; visual and audio facilities.

CHAPTER 8 / GRAPHICS AND VISUAL AIDS

DO

- make liberal, but proper, use of figures and tables
- use descriptive titles for figures and tables
- keep visual aids as simple as possible
- carefully select the scale of a histogram
- use visual aids in almost all oral presentations

DON'T

- forget to include the units of variables in figures
- use colored figures in written reports

8.1 INTRODUCTION

We have all heard the saying, "A picture is worth a thousand words." This is very true in written and oral communication, but only if the illustration is well done. A poorly composed illustration, whether a picture, graph, or a table, may be worth much less than the proverbial one thousand words and, in some cases, may actually detract from the presentation. Imagine how successful a professional would be if, while making an oral progress report to an important client, he or she used visual aids that could not be read by those in the meeting room. Imagine the confusion that would be created if a professional included in a final project report tables with inadequately labeled column headings or tables that did not specify the units of the values in the columns. The quality of graphics and visual aids is important to the success of a communication project.

Tables and illustrations are vital parts of both written and oral communication. They summarize important results and help show the effects of variables. They put the material in an easily understood form; tables and figures are often easier to comprehend than prose or an oral presentation. Visual aids serve the same purpose in a speech as tables or figures do in a written report. Both increase the audience's comprehension of the material and add variety. Visual aids can also serve as an outline so the speaker can avoid using notes. The objective of this chapter is to introduce the basics for formulating and preparing tables, figures, and visual aids.

8.2   TABLES

Tabular data are common and important in almost all reports in professional life, from engineering and science to business and medical research. Tables are used to present tabular data including columns of related numbers (see Table 8), columns of qualitative information such as descriptions of classification requirements (see Table 9), or columns of mixed information such as equations and statistical criteria (see Table 10). While there are no constraints on the structure of a table, a table will best communicate if the information is placed in a systematic form; tables listing unrelated information may not be effective.

The format of a table is important. Each table should have a descriptive title, which appears at the top of the table preceded by the word TABLE, in capitals, and a table number. For example:

TABLE 33. Summary Statistics for the Evaporation Data.

Since every table should be easily understood when apart from the text, the title should explicitly identify the content. The title can also include the definition of notation used in the table and the units of the variables, if these are not included as part of the column headings. For example, the title, Computation of Nonparametric Correlation Coefficients, would not be an adequate title for the information shown in Table 8. As shown on the table, the title explains the content, the specific method used, and some of the notation. While format practices are not totally uniform, word titles are commonly capitalized but first-letter capitals used for the re-mainder of the title.

Column headings can also communicate important information. In general, the headings describe how the information or data are being distinguished. For example, in Table 9a, the column headings indicate that earth dams are classified by storage volume and height. In Table 9b, headings suggest that loss of life and economic loss are criteria for iden-tifying the hazard potential. To maintain proper column spacing, column headings often include abbreviations, with the abbreviations defined in either the table title or footnote. For example, both the peak discharge (Q_p) and the correlation coefficient (r_s) are defined in the title of Table 8, while the difference d_i is defined in the footnote. The units of Q_p, which are enclosed within parentheses, are included as part of the column head-ing in Table 8. In some cases, the column headings are underlined, al-ternatively, a solid line across the table width is used. Tables with a large number of columns and text references to specific columns should include column numbers (see Table 11).

TABLE 8. Calculation of Spearman Correlation Coefficients (r_s) between Peak Discharge (Q_p) and Rainfall Characteristics (D, V, and I)

Year	Q_p (ft³/sec)	Rank	Storm Duration D D (hr)	Rank	d_i	Total Volume V V (in.)	Rank	d_i	Maximum Intensity I I (in./hr)	Rank	d_i
1945	2000	14	21	4.5	−9.5	3.02	5	−9.0	0.37	16.5	2.5
1946	1740	15	13	11	−4.0	1.93	13	−2.0	0.29	19	4.0
1948	2060	13	10	15	2.0	1.89	14	1.0	0.64	9	−4.0
1949	1530	19	8	20	1.0	0.73	24	5.0	0.32	18	−1.0
1950	1600	18	17	8	−10.0	3.97	3	−15.0	1.37	2	−16.0
1951	1690	16	21	4.5	−11.5	2.49	7	−9.0	0.42	15	−1.0
1952	1420	21	23	3	−18	2.36	8	−13.0	0.60	10.5	−1.05
1953	1330	23	20	6	−17	1.87	15	−8.0	0.23	23	0.0
1954	607	24	4	23.5	−0.5	0.80	23	−1.0	0.28	21	−3.0
1955	1380	22	9	18	−4.0	0.82	22	0.0	0.19	24	2.0
1956	1660	17	24	2	−15.0	2.19	9	−8.0	0.37	16.5	−0.5
1957	2290	12	15	9.5	−2.5	2.17	10	−2.0	0.28	21	9.0
1958	2590	9	15	9.5	0.5	2.90	6	−3.0	0.74	7	−2.0
1959	3260	6	11	13	7.0	2.12	11	5.0	0.60	10.5	4.5
1960	2490	11	10	15	4.0	5.04	2	−9.0	1.43	1	−10.0
1961	2080	8	19	7	−1.0	3.12	4	−4.0	0.78	5	−3.0
1962	2520	10	12	12	2.0	1.64	16.5	6.5	0.56	12	2.0
1963	3360	5	7	21.5	16.5	1.64	16.5	11.5	0.48	13	8.0
1964	8020	1	25	1	0.0	6.98	1	0.0	0.77	6	5.0
1965	4310	4	9	18	14.0	1.52	20	16.0	0.66	8	4.0
1966	4380	2	9	18	16.0	1.59	18	16.0	0.47	14	12.0
1967	3220	7	4	23.5	16.5	1.42	21	14.0	1.16	3	−4.0
1968	4320	3	7	21.5	18.5	2.07	12	9.0	0.97	4	1.0
d_i^2					2740			1813			905
r_s					−0.19			0.21			0.61

Note: d_i = the difference between the rank of the rainfall characteristics (D, V, or I) and the rank of the peak discharge (Q_p)

TABLE 9. Corps of Engineers' Classification System for Earth Dams

(a) Size Classification

| Category | Impoundment | |
	STORAGE (ac-ft)	HEIGHT (ft)
Small	<1,000 and ≥50	<40 and ≥25
Intermediate	≥1,000 and <50,000	≤40 and <100
Large	≥50,000	≥100

(b) Hazard Potential Classification

Category	Loss of Life (Extent of Development)	Economic Loss (Extent of Development)
Low	None expected (no permanent structures for human habitational)	Minimal (undeveloped to occasional structures or agricultural)
Significance	Low (no urban developments and no more than a small number of inhabitable structures)	Appreciable (notable agriculture industry, or structures)
High	More than few	Excessive (extensive community, industry, or agriculture)

Ref: "Recommended Guidelines for Safety Inspection of Dams." (1975). *Appendix D*, U.S. Army Corps of Engineers National Prgm of Inspection of Dams, Washington, D.C.

Numerical values shown in table columns should be properly aligned. For columns of integer values, the numbers should be right justified. For numbers with decimal points, the values should be aligned with the decimal points. Numbers should be shown using commonly accepted rules for specifying the number of significant digits. Computer programs that print several unnecessary trailing zeros should be modified to eliminate the meaningless zeros. Numbers less than 1 in absolute value should include a leading zero, for example, 0.023 rather than .023.

Footnotes can be an important part of tables. They can be used to indicate the source of the table (e.g., Table 9), provide notation and def-

TABLE 10. Summary of Hypothesis Tests

H_0	Test Statistic	H_A	Region of Rejection
$\mu = \mu_o$ (σ *known*)	$Z = \dfrac{\overline{X} - \mu}{\sigma/\sqrt{n}}$	$\mu < \mu_o$ $\mu > \mu_o$ $\mu \neq \mu_o$	$Z < -Z_\alpha$ $Z > Z_\alpha$ $\left\{ \begin{array}{l} Z < -Z_{\alpha/2} \\ and \\ Z > Z_{\alpha/2} \end{array} \right.$
$\mu = \mu_o$ (σ *unknown*)	$t = \dfrac{\overline{X} - \mu_o}{s/\sqrt{n}}$ $\nu = n - 1$	$\mu < \mu_o$ $\mu > \mu_o$ $\mu \neq \mu_o$	$t < -t_\alpha$ $t > t_\alpha$ $\left\{ \begin{array}{l} t < -t_{\alpha/2} \\ and \\ t > t_{\alpha/2} \end{array} \right.$
$\sigma^2 = \sigma_o^2$	$\chi^2 = \dfrac{(n-1)\,S^2}{\sigma^2}$ $\nu = n - 1$	$\sigma^2 < \sigma_o^2$ $\sigma^2 > \sigma_o^2$ $\sigma^2 \neq \sigma_o^2$	$\chi^2 < -\chi^2_{1-\alpha}$ $\chi^2 > \chi^2_\alpha$ $\left\{ \begin{array}{l} \chi^2 < \chi^2_{1-\alpha/2} \\ and \\ \chi^2 > \chi^2_{\alpha/2} \end{array} \right.$

initions (e.g., see Table 8), or to explain information in the table. Footnotes should be placed below a blank line of separation. A reference mark is sometimes used to indicate the column or part of the table that the footnote refers to.

Tables should be located as close as possible to the place in the text where they are referenced. In books and reports, the table often appears on the same page or adjacent page as the point of reference. For some types of writing, each table is placed on a separate page immediately following the first or primary reference to it. If the width of a table is greater than its length, it can be placed sideways in the report, with the title on the margin adjacent to the binding. Table 8 illustrates this.

Each table should be numbered. In some writings, the number includes the chapter number as part of the table number. For a report that includes only a few tables and does not have chapters or sections, the table number would be indicated by an Arabic number. Tables should be numbered consecutively in the report or in the chapter when the table number includes the chapter number. When referencing a table, place the word "Table" in first-letter capital form, followed by the table number, for example, Table 6, Table II-4, or Table 7-3.

Tabular data are not frequently used as visual aids for oral reports. Unless the audience is small and the table very simple, it is difficult to make a table that can be seen by everyone in the audience. For example, Table 8 contains far too much detail to be an effective visual aid for an

TABLE 11.   Adjustment of Rubio Wash Annual Maximum Flood Record for Urbanization

(1)	(2)	(3)	(4)	(5)	(6)	(7)	(8)	(9)	(10)	(11)	(12)	(13)	(14)
1929	18	661	47	0.959	1.55	2.06	878	47	0.959	1.55	2.06	878	47
1930	18	1,690	30	0.612	1.43	1.84	2,175	22	0.449	1.39	1.77	2,152	22
1931	19	798	46	0.939	1.54	2.03	1,052	44	0.898	1.53	1.99	1,038	44
1932	20	1,510	34	0.694	1.50	1.87	1,882	32	0.653	1.49	1.85	1,875	32
1933	20	2,070	20	0.408	1.44	1.75	2,516	14	0.286	1.40	1.70	2,514	14
1934	21	1,680	31	0.633	1.52	1.85	2,045	25	0.510	1.49	1.80	2,030	28
1935	21	1,370	35	0.714	1.54	1.87	1,664	34	0.694	1.53	1.87	1,664	34
1936	22	1,180	40	0.816	1.59	1.89	1,403	36	0.735	1.55	1.88	1,431	36
1937	23	2,400	14	0.286	1.45	1.70	2,814	9	0.184	1.42	1.65	2,789	9
1938	25	1,720	29	0.592	1.55	1.83	2,031	28	0.571	1.55	1.83	2,031	27
1939	26	1,000	43	0.878	1.67	1.97	1,180	42	0.857	1.66	1.95	1,175	42
1940	28	1,940	26	0.531	1.58	1.80	2,210	20	0.408	1.55	1.75	2,190	20
1941	29	1,200	38	0.776	1.69	1.91	1,356	37	0.755	1.68	1.90	1,357	37
1942	30	2,780	8	0.163	1.50	1.64	3,039	5	0.102	1.47	1.60	3,026	5
1943	31	1,930	27	0.551	1.68	1.81	2,079	23	0.469	1.64	1.78	2,095	23
1944	33	1,780	28	0.571	1.70	1.82	1,906	31	0.633	1.73	1.84	1,893	31
1945	34	1,630	32	0.653	1.75	1.85	1,723	33	0.673	1.76	1.86	1,723	33
1946	34	2,650	10	0.204	1.59	1.67	2,783	10	0.204	1.59	1.67	2,783	10
1947	35	2,090	19	0.388	1.67	1.75	2,190	21	0.429	1.69	1.77	2,189	21
1948	36	530	48	0.980	2.02	2.12	556	48	0.980	2.02	2.12	556	48
1949	37	1,060	42	0.857	1.90	1.95	1,088	43	0.816	1.87	1.92	1,088	43
1950	38	2,290	17	0.347	1.69	1.74	2,358	16	0.327	1.68	1.73	2,358	16
1951	38	3,020	4	0.082	1.55	1.57	3,059	4	0.082	1.55	1.57	3,059	4
1952	39	2,200	18	0.367	1.72	1.75	2,238	19	0.388	1.72	1.75	2,238	19
1953	39	2,310	15	0.306	1.70	1.71	2,324	17	0.347	1.71	1.73	2,337	17
1954	39	1,290	36	0.735	1.86	1.88	1,304	38	0.776	1.88	1.90	1,304	38
1955	39	1,970	25	0.510	1.78	1.80	1,992	29	0.592	1.80	1.82	1,992	29
1956	39	2,980	5	0.102	1.58	1.60	3,018	6	0.122	1.58	1.60	3,018	6

TABLE 11. Continued

(1)	(2)	(3)	(4)	(5)	(6)	(7)	(8)	(9)	(10)	(11)	(12)	(13)	(14)
1957	39	1,740	9	0.184	1.63	1.65	2,774	11	0.224	1.65	1.67	2,773	11
1958	39	2,780	7	0.143	1.60	1.62	2,815	8	0.163	1.62	1.64	2,814	8
1959	39	985	44	0.898	1.96	1.99	1,000	45	0.918	1.99	2.01	995	45
1960	39	902	45	0.918	1.99	2.01	911	46	0.939	2.01	2.03	911	46
1961	39	1,200	39	0.760	1.88	1.90	1,213	40	0.816	1.91	1.93	1,213	40
1962	39	1,180	41	0.837	1.93	1.95	1,192	41	0.837	1.93	1.95	1,192	41
1963	39	1,570	33	0.673	1.83	1.85	1,587	36	0.714	1.85	1.87	1,587	35
1964	40	2,040	23	0.469			2,040	27	0.551			2,040	26
1965	40	2,300	16	0.327			2,300	18	0.367			2,300	18
1966	40	2,040	22	0.449			2,040	26	0.531			2,040	25
1967	40	2,460	13	0.265			2,460	15	0.306			2,460	15
1968	40	2,890	6	0.122			2,890	7	0.143			2,890	7
1969	40	2,540	12	0.245			2,540	13	0.265			2,540	13
1970	40	3,700	1	0.020			3,700	1	0.020			3,700	1
1971	40	1,240	37	0.755			1,240	39	0.796			1,240	39
1972	40	3,166	3	0.061			3,166	3	0.061			3,166	3
1973	40	1,985	24	0.490			1,985	30	0.612			1,985	30
1974	40	3,180	2	0.041			3,180	2	0.041			3,180	2
1975	40	2,070	21	0.429			2,070	24	0.490			2,070	24
1976	40	1,610	11	0.224			2,610	12	0.245			2,610	12

(1) = water year.
(2) = percentage of imperviousness.
(3) = measured annual maximum discharge.
(4) = rank of events in (3).
(5) = Weibull exceedence probability for rank in (4).
(6) = adjustment factor f_1.
(7) = adjustment factor f_2.
(8) = adjusted annual maximum.
(9) = rank of events in (8).
(10) = Weibull exceedence probability.
(11) = adjustment factor f_1.
(12) = adjustment factor f_2.
(13) = adjusted annual maximum.
(14) = rank of events in (13).

oral presentation. Table 9 is simple enough that it may make an acceptable slide or transparency for use with a small audience. Where it is necessary to use tabular form, the table should be presented so that only the necessary information is shown. Quantities that will not be specifically referenced should not be included on the slide or transparency.

A few other points on tables: (1) Tables can have multiple parts, such as Table 9; (2) tables can have layered column headings, such as those shown in Table 8; and (3) tables requiring more than two or three pages or multiple tables illustrating the same point should be placed in an appendix so they don't distract the reader. Most will not want to read all of the quantities in an extensive table. When a table must be continued on a second page, write "continued" at the bottom of the first page of the table, and as the title for the second page, indicate the table number followed by the word "continued" in parentheses.

8.3 ILLUSTRATIONS

All illustrations except tabular data should be labeled figures. Many of the rules for tables are applicable to illustrations. The categories of illustration can include figures (e.g., graphs and sketches), nomographs, photographs, or descriptive summaries not in tabular form. As with tables, each illustration should be placed on a separate page immediately following the first reference to the illustration. Figures should be designated by an Arabic numeral and numbered consecutively throughout the report. For reports that are subdivided by sections or chapters, the section or chapter number may be included as part of the figure number. On the figure itself and when referring to the figure, the word "figure" should have the first letter capitalized, that is Fig. 22. This is also an example showing the format of a figure.

In addition to being numbered, each figure requires a descriptive title, placed at the bottom of the figure using first-letter capitals. Like tables, figure titles can be used to describe the content of the figure, define notation, and specify units of variables, if the units are not specified on the axes.

The axes of a figure should also be labeled, including the units of the variables (see Fig. 22). For figures with multiple lines, a description of the individual lines should be placed next to the line, unless it is part of the title or included in a box within the figure itself. Fig. 23 shows an example of a graph that involves three variables, with the values of the third variable (B) placed next to the respective curve.

Except in special cases, illustrations should be black and white. Colored lines should not be used because they are indistinguishable on photocopies. For multiple-line illustrations, various combinations of broken lines can be used to distinguish them. A few such forms are: (——————),

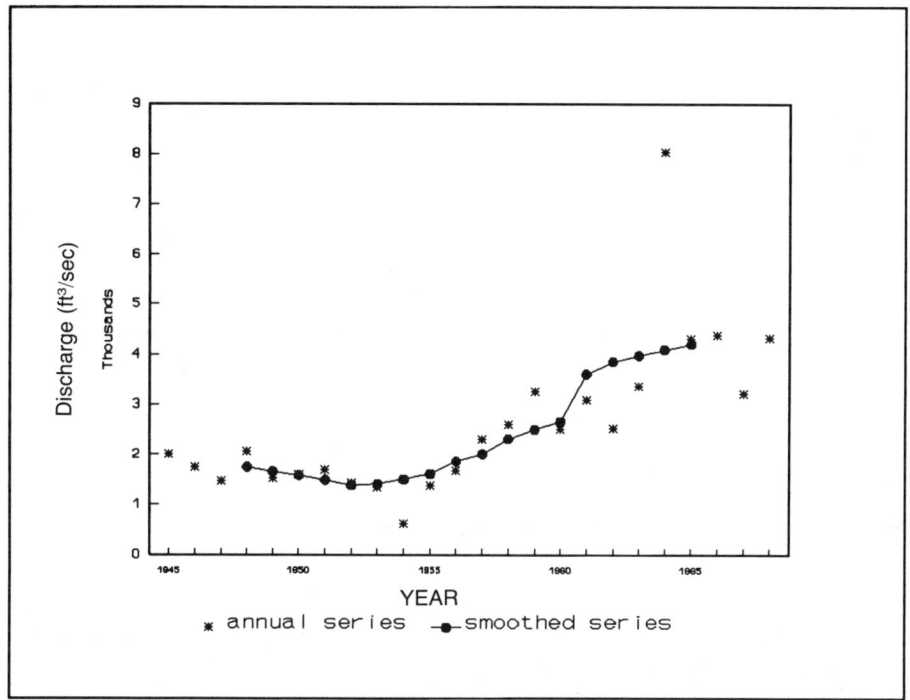

FIG. 22.—*Annual Flood Series and Smoothed Series for Pond Creek Watershed, 1945–1968.*

(————), (- - - - -), (–•–•), (–••–). Fig. 24 shows an example of lines for four different ranks indicated by lines of different weight.

Hand lettering is acceptable only when there is no alternative and it must be expertly done. With the availability of graphics packages on personal computers, it is rarely necessary to use letter by hand. Folded illustrations should be avoided since they are troublesome to reproduce and tend to deteriorate much more quickly than the remainder of the report.

Figs. 22, 23, and 24 are typical of technical reports in engineering; they are simple x-y plots that can show measured data (asterisks in Fig. 22), the effect of a third variable (B in Fig. 23), or a trend (solid line in Fig. 22). Other types of data can be presented in figures. Fig. 25 shows the time trend of helmet use in Texas before and after the passage of a law requiring cyclists to wear helmets. The figure of a cyclist and the outline of Texas add variety and give greater meaning to the figure; the extras should enhance the viewer's interest in looking at the figure, as well as make a longer-lasting impression. Fig. 26 also shows a before-and-

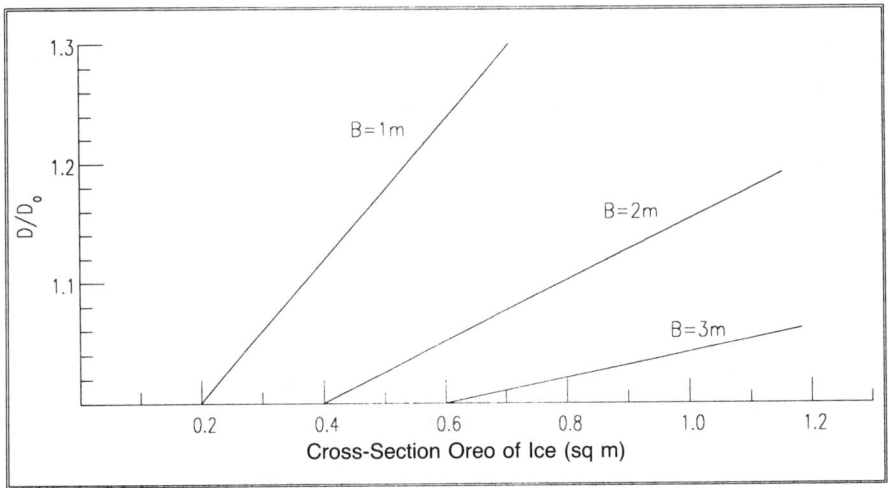

FIG. 23.—*Graphical Presentation of Three-Variable Relationship*

after comparison, using the human form to add variety and enhance comprehension of the data presented.

8.3.2   Pie Charts

When data are expressed as percentages, proportions, or fractions of a whole, pie charts can be used to enhance the material. A round circle is used to represent 100% and the "pie" is sectioned according to the percentages. Figure 27 shows of a pie chart of drivers who use radar detectors. The same material could be presented as a table but it would not be as effective. The pie chart has the advantage that the size of the pie slice supports the numerical values. In Figure 27, the width of the pie is used to show a second factor, that is 56 percent of the drivers who use radar detectors drive faster.

8.3.3   Histograms

Histograms are also an effective form of graphics (see section 4.2.1). The lengths of the bars in the histogram reflect the magnitude. The histogram is especially effective when the exact numerical value is less than the differences between the ordinates. Fig. 28 shows a histogram of the base price of five automobiles. In Fig. 28a the scale for the ordinate is limited to the range from $10,000 to $14,000; this gives the impression of a large difference between the smallest and largest ordinates. In Fig. 28b, the scale for the ordinate is given from $0 to $14,000, so the differences between the ordinates are perceived to be relatively small (compared with part a). As a general rule, the scale should be selected to present the data

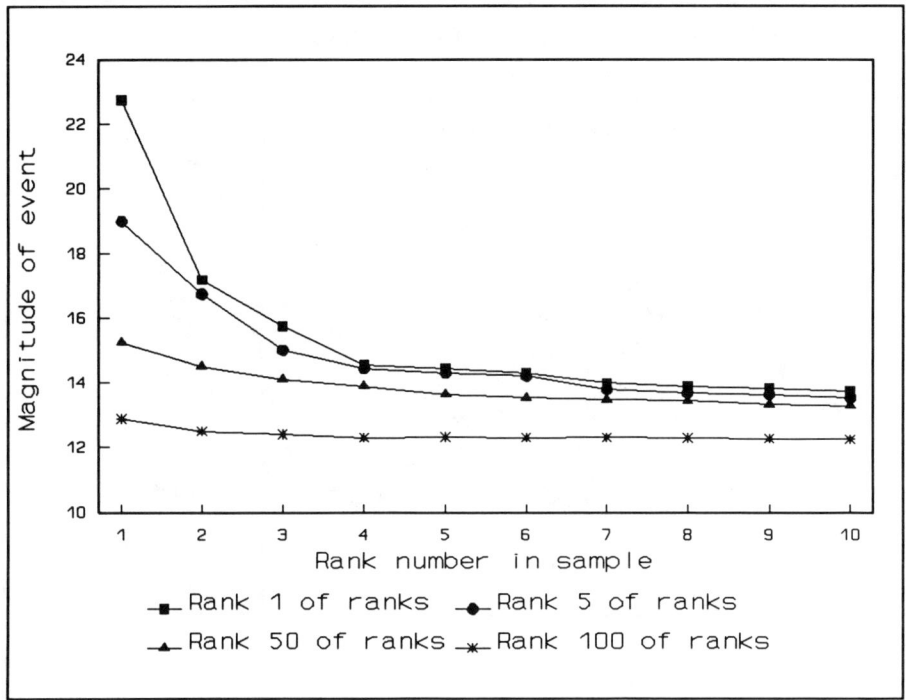

FIG. 24.—*Rank Numbers for Highest Sample Ranks*

in such a way that meaningful differences are evident and insignificant differences do not appear as differences. Unfortunately, histograms are sometimes structured to support the author's bias by making a difference appear significant when the difference is not really important.

Histograms can also be used when more than one variable is involved. In such cases, the variables are grouped together for each item. Fig. 29 characterizes the nutritional content of five breakfast cereals by showing three ingredients of the cereals: fat, fiber, and sugar. The histogram clearly shows the differences in the nutritional content of the five types of breakfast cereal.

8.4 VISUAL AIDS

Visual aids can make or break an oral presentation. There are a number of visual aids (VA's) that can be used to improve a presentation. The three most commonly used VA's are handouts, transparencies, and 35-mm slides. Each has advantages and disadvantages. A photocopied hand-out is inexpensive, can be done at the last moment, and provides some-

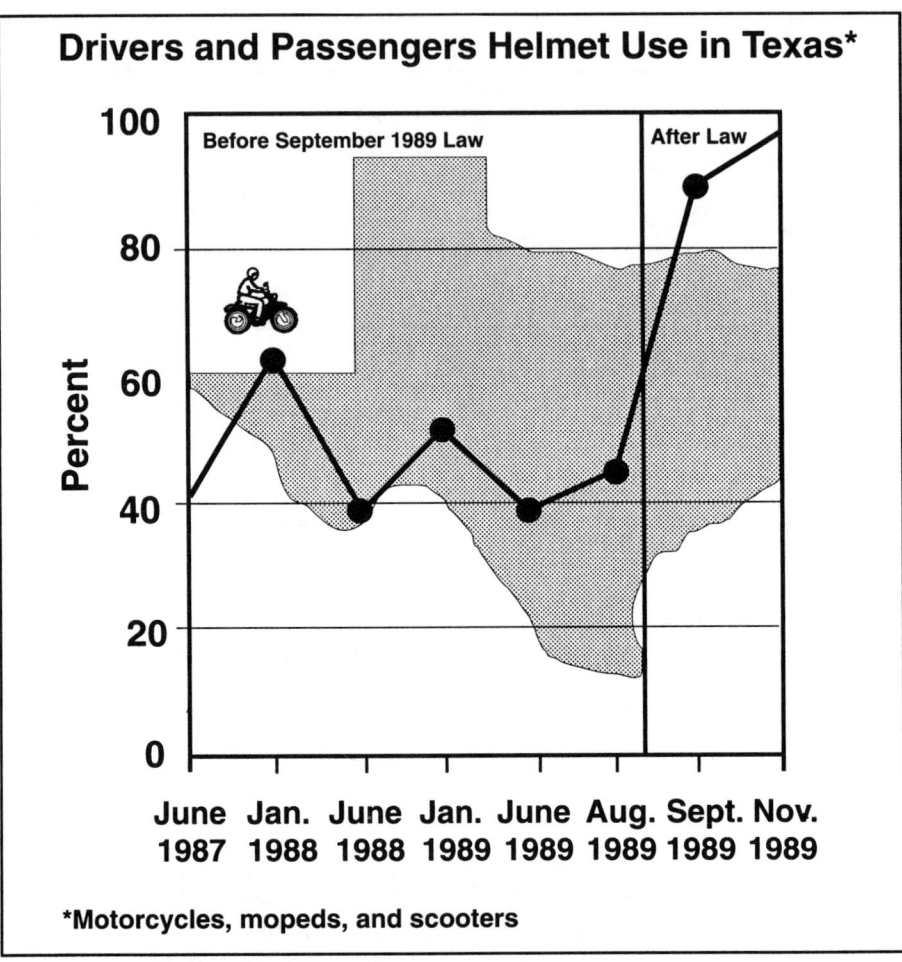

FIG. 25.—*Graphical-Pictorial Communication of Data (Ref: Insurance Institute for Highway Safety Status Report (1990). Vol. 25(4), Apr. 7.)*

thing that can be taken from the presentation. However, handouts can be distracting if the audience pages through them at a pace different from presentation. Handouts also must be distributed.

Properly prepared 35 mm slides are the preferred visual aid for large audiences or where some measure of sophistication is desired. However, they are the most costly aid, require both camera equipment (unless they are commercially prepared) and one or more days of preparation time.

Transparencies are a convenient alternative to handouts and slides. They are inexpensive and can be quickly and conveniently prepared from a photocopy. They require only a screen (or a white wall) and an overhead

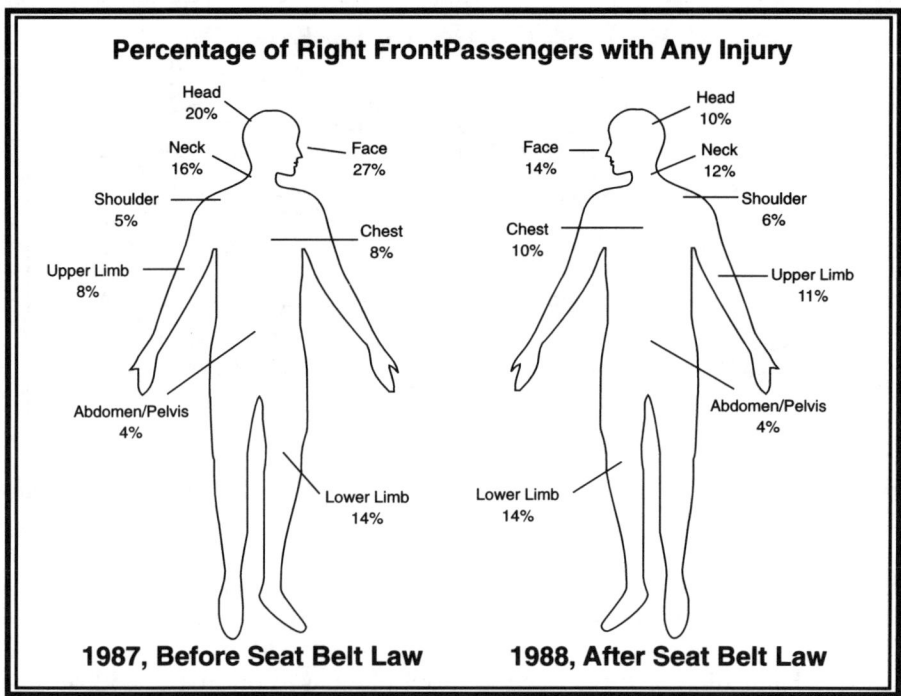

Percentage of Right FrontPassengers with Any Injury

Head
20%

Neck
16%

Face
27%

Shoulder
5%

Chest
8%

Upper Limb
8%

Abdomen/Pelvis
4%

Lower Limb
14%

Head
10%

Face
14%

Neck
12%

Shoulder
6%

Chest
10%

Upper Limb
11%

Abdomen/Pelvis
4%

Lower Limb
14%

1987, Before Seat Belt Law   **1988, After Seat Belt Law**

FIG. 26.—*Pictorial Communication of Data (Ref: Insurance Institute for Highway Safety Status Report, (1990) Vol. 25(1), Jan. 27, 1990)*

projector. They do not require a completely dark room and they allow the speaker to move around. However, the overhead projector can be a distraction since the audience sometimes must look around the projector. The shuffling and replacing of transparencies can also be rather distracting. If transparencies are used, the speaker should take special precautions to ensure that they are maintained in an orderly fashion.

Visual aids offer many advantages. They can significantly improve the quality of a presentation, but only when they are well prepared. The following points should be considered when preparing and using transparencies (many of these points also apply to 35-mm slides):

1. The projected area of an overhead projector is square, so when preparing the transparencies, position the material in a square format, not one or two lines across the page. This is illustrated in the two parts of Fig. 30. The material in Fig. 30a is expanded to fill the same space as that of Fig. 30b. Because Fig. 30b places

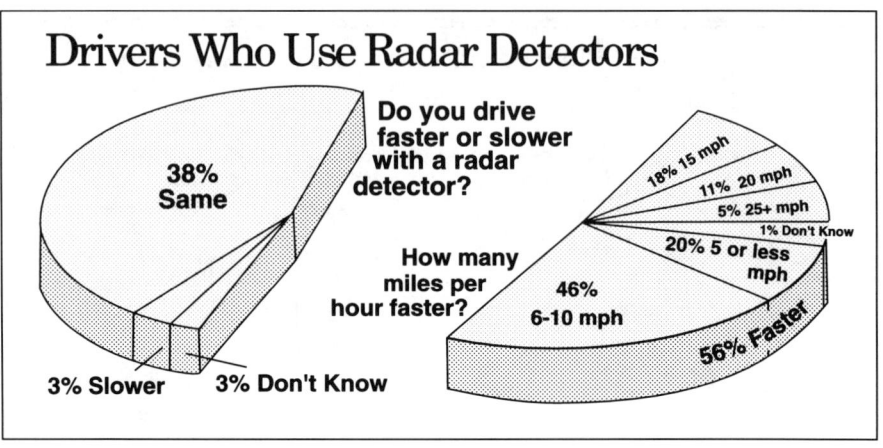

FIG. 27.—*Example of Pie Chart (Ref: Insurance Institute for Highway Safety Status Report, Vol. 23(11), Dec. 3, 1988)*

the material in a square format, the print can be enlarged on a photocopier so that it is much easier to read.

2. Each slide or transparency should be based on just one idea; this will enable it to be simple, yet forceful.

3. Use outline form (rather than complete sentences). As a rule-of-thumb, limit each transparency to 15 or 20 words. This keeps the transparency simple, yet ensures that the focus is on the major point of the transparency. A comparison of the material in Figs. 31a and 31b suggests that outline form is easier to comprehend and when enlarged, easier to read.

4. Avoid detailed tables or figures. Too much information will be difficult to read and the minds of those in the audience will tend to wander from the material being emphasized by the speaker. Figure 32 shows a table that would be virtually impossible to read as a transparency.

5. Make sure the transparencies are in proper order. This will eliminate disruptions when a transparency cannot be readily located. Disorganization of visuals will also reduce the confidence of the audience in the speaker. When using 35-mm slides, go through the slides prior to the presentation to ensure that they are in the proper order and not upside down.

6. Place a blank sheet of paper between each pair of transparencies to prevent the transparencies from sticking together and to facilitate handling. The clean sheet of paper also enables the speaker

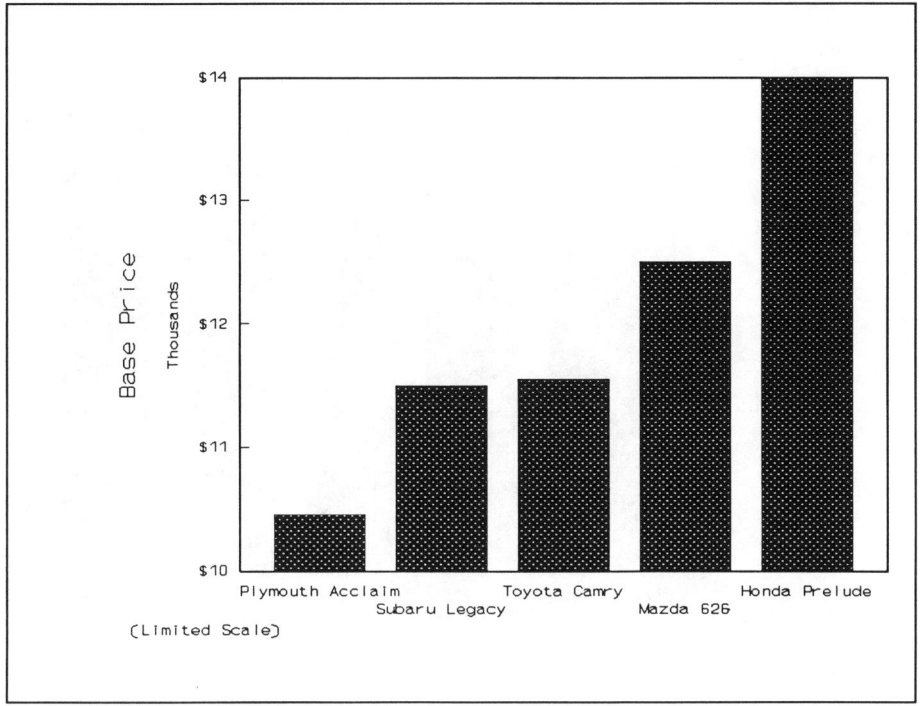

FIG. 28a.—*Base Price of Selected 1990 Automobiles: Limited Scale*

to read the transparency prior to placing it on the overhead projector.

7. Have duplicate transparencies available, and in sequence, if you intend to repeat the material on a transparency (such as the objectives of the presentation). Don't search through the transparencies already used for the one discussed previously.

8. Use boldface for the item to be discussed (see Fig. 33). This motivates the audience to focus on the idea being discussed.

9. Take a grease pencil or transparency marker to the presentation to emphasize material on a transparency. Also some unused transparencies will be handy when responding to questions.

10. Do not block the view of the audience during the presentation.

11. When placing a transparency on the projector, quickly glance at the screen to ensure that the transparency is properly positioned. But make sure that you do not continue to look at and talk to the screen.

12. If a transparency includes a number of unrelated items, such as

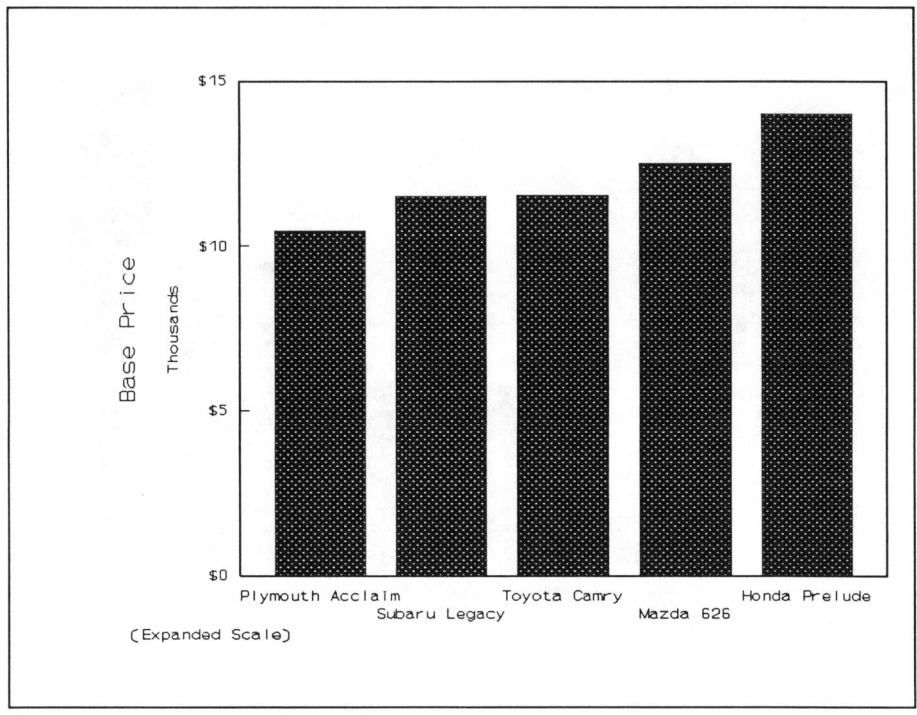

FIG. 28b.—*Base Price of Selected 1990 Automobiles: Expanded Scale*

a list of objectives or a list of conclusions, use a piece of paper to conceal those that have not been discussed. Move the paper down the transparency as the items are discussed. This helps to ensure that the audience will focus on the point that you are discussing. Fig. 33 shows an alternative way of presenting a list of items; however this approach requires multiple slides or transparencies.

13. If you are using transparencies occasionally during your presentation, turn off the projector between examples. This will help focus the audience on you, not the screen.

14. Simplify figures adapted from a written report. As shown in Fig. 34, the visual aid for an oral presentation (part b) is less detailed than the same figure in the written report (part a).

15. Do not use vertical lettering such as the label on the ordinate of Fig. 34a; instead, simplify it and use a horizontal format, as shown in Fig. 34b.

16. Visual aids can also be too simple. Fig. 35 shows a slide does not

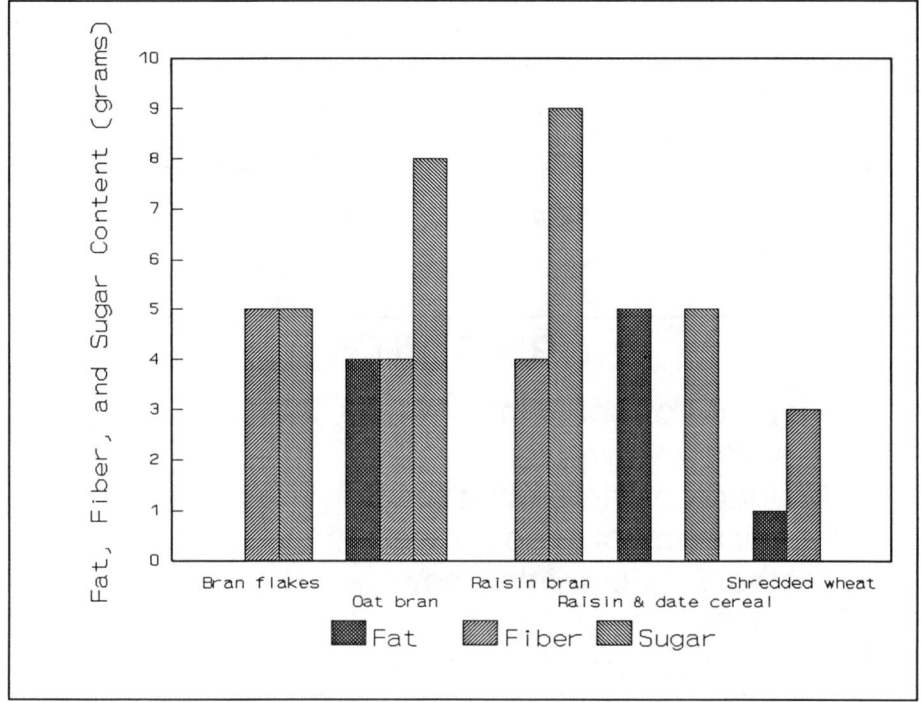

FIG. 29.—*Nutrition Content of Selected Breakfast Cereals (Fat = F; Fiber = I; Sugar = S)*

communicate information. While Fig. 33 could be simplified, Fig. 35 is too simple.
17. The spacing should be sufficient to be read easily from the most distant part of the room. A double-spaced format is usually adequate.

8.5 COMPUTER GRAPHICS AND DRAWINGS

Computer software has advanced to the point where it is now simple to create sketches, graphs, and other professionally drawn figures on a personal computer. There is a wide variety of graphics packages available. The best choice of package will depend on the type of computer being used, the software environment within the computer (e.g., Windows), and the memory available.

There are many reasons to use computer graphics in a formal report. When printed on a high quality printer, computer graphics can have a very professional appearance. It is easy to obtain a high degree of accuracy when plotting points on a computer-generated graph; therefore, the illustration is professional in appearance, as well as in technical content.

CREATIVITY

1. Methods of obtaining a fresh viewpoint
2. Methods of increasing the efficiency of problem-solving
3. Methods of obtaining new answers to old problems
4. Imaginative problem-solving

FIG. 30a.—*Transparency Material in Linear Form*

CREATIVITY
1. Methods of obtaining
   a fresh viewpoint
2. Methods of increasing the
   efficiency of problem-solving
3. Methods of obtaining new
   answers to old problems
4. Imaginative problem-solving

FIG. 30b.—*Transparency Material in Block Form*

Another advantage of computer graphics over drafted drawings is that a computer-generated graphic can be easily updated or revised. The drawing can be linked to the data source so that when the data file is updated, the drawing is automatically updated. Alternatively, if you choose to change some object on a sketch or change an axis label on a plot, this revision will take a matter of minutes on a computer (assuming that the previous graph was saved!) as opposed to hours required to redraft the figure.

The only disadvantage to using computer graphics software is that you have to learn how to use the software. There is certainly a learning curve associated with the use of these powerful tools; however, once mastered, the advantages far outweigh this disadvantage.

In choosing the appropriate graphics package, consider where your data for any plots will come from. Many graphics packages can import data from spreadsheets or from portions of spreadsheets. You should never need to retype data; make sure that your graphics packages can access the data either from anywhere in a spreadsheet or from a data file,

```
┌─────────────────────────────────────────┐
│            CREATIVITY                    │
│                                          │
│   1. SYNECTICS                           │
│      Focus on underlying concepts        │
│   2. BRAINSTORMING                       │
│      Focus on specific problem           │
│   3. CHECKLIST                           │
│      Focus on idea improvement           │
│                                          │
└─────────────────────────────────────────┘
```

FIG. 31a.—*Outline Form*

```
┌───────────────────────────────────────────────────────────────┐
│                    CREATIVITY STIMULATORS                      │
│                                                                │
│  1. SYNECTICS: Synectics is a method that focuses on the        │
│     underlying concept.                                        │
│  2. BRAINSTORMING: Brainstorming is a method that focuses on    │
│     a specific problem.                                        │
│  3. CHECKLIST: The checklist approach focuses on the           │
│     improvement of an existing idea.                           │
│                                                                │
└───────────────────────────────────────────────────────────────┘
```

FIG. 31b.—*Sentence Form*

such as an ASCII file. In addition, check the software for flexibility. Will you be able to change the sizes of objects and lettering? Can you draw freehand with the aid of a mouse? Can you move objects from one point to another with relative ease? Is the software compatible with your word processor so that the figures that you have created can be imported into a document?

One additional consideration might be compatibility with graphics packages being used at work or school. If you wish to work with your figures both at work or school as well as on your personal computer at home, you will want to choose the same graphics software.

If you choose a flexible package, there will be a few limitations on the drawings, figures, and graphs that you can produce. One might be in the text appearance, again depending on the software chosen. For example, some packages are not able to produce subscripts or superscripts within the text or generate mathematical symbols. These limitations can often be overcome by importing the figure into a word processor and producing the text there.

Learning to create drawings using computer software is not difficult. The best way to learn is to ask someone (a friend or professor) to show you how to get started. It will take only an hour or two of

TABLE X.　Percentage Removal of Total Solids due to Various Street Sweeping Operations (Area 3: Multi-Family Residential Land Use)

Type of Street Sweeper	Pollutant	Forward speed of street sweeper											
		3.0 mph						4.0 mph					
	Frequency of Sweeping	2.0 Days	3.0 Days	4.0 Days	5.0 Days	6.0 Days	7.0 Days	2.0 Days	3.0 Days	4.0 Days	5.0 Days	6.0 Days	7.0 Days
Vacuumized	Total Solids	69.1	60.2	53.0	47.3	43.2	37.9	68.0	59.1	52.1	46.6	42.6	37.4
	Volatile Solids	57.5	48.0	41.4	36.6	33.5	29.5	55.1	46.1	39.8	35.4	32.4	28.7
	BOD	55.9	46.3	39.9	35.2	32.2	28.3	53.4	44.3	38.2	33.9	31.0	27.4
	COD	44.3	33.3	26.4	21.9	19.5	16.8	40.5	30.0	23.6	19.6	17.6	15.3
	Kjeldahl Nitrogen	57.4	47.6	40.7	35.7	32.5	28.5	55.0	45.6	39.0	34.4	31.4	27.6
	Nitrates	46.8	36.5	39.9	25.5	22.9	19.9	43.3	33.6	27.5	23.6	21.3	18.7
	Phosphates	33.1	22.4	16.2	12.7	11.2	9.7	28.3	18.5	13.1	10.3	9.1	8.1
	Total Heavy Metals	63.4	54.1	47.3	42.0	38.3	33.6	61.6	52.5	45.9	40.9	37.3	32.8
Motorized	Total Solids	68.7	59.7	52.7	47.0	43.0	37.8	67.6	58.7	51.8	46.3	42.4	37.3
	Volatile Solids	57.5	47.9	41.3	36.5	33.5	29.5	55.2	46.0	39.7	35.2	32.3	28.6
	BOD	55.9	46.3	39.8	35.1	32.1	29.3	53.4	44.2	38.1	33.7	30.0	27.3
	COD	45.3	34.2	27.2	22.6	20.1	17.3	41.7	31.0	24.4	20.3	18.2	15.8
	Kjeldahl Nitrogen	57.4	47.6	40.7	35.7	32.5	28.6	55.2	45.6	39.0	34.4	31.4	27.7
	Nitrates	47.7	37.3	30.6	26.0	23.4	20.3	44.4	34.4	28.2	24.1	21.7	19.0
	Phosphates	34.8	23.8	17.4	13.7	12.1	10.4	30.3	20.1	14.4	11.3	10.0	8.8
	Total Heavy Metals	63.1	53.8	47.0	41.8	38.1	33.4	61.3	52.2	45.6	40.6	37.1	32.7

FIG. 32.—*Table too Detailed for Audio-Visual*

CONCLUSIONS

1. **DETENTION BASINS CAN INCREASE DOWNSTREAM EROSION WHEN THEY ARE DESIGNED FOR FLOOD CONTROL**
2.
3.

CONCLUSIONS

1. Detention basins can increase downstream erosion when they are designed for flood control
2. **DETENTION BASINS DESIGNED FOR EROSION CONTROL REQUIRE MORE STORAGE THAN BASINS DESIGNED FOR FLOOD CONTROL**
3.

CONCLUSIONS

1. Detention basins can increase downstream erosion when they are designed for flood control
2. Detention basins designed for erosion control require more storage than basins designed for flood control
3. **THE EFFICIENCY OF DETENTION BASINS DESIGNED FOR EROSION CONTROL DEPENDS ON LENGTH/DEPTH CHARACTERISTICS**

FIG. 33.—*Successive Introduction of List Items*

experimentation (and occasional frustration!) to learn how to make professional-looking figures. If you don't know anyone who can help you, you will have to rely entirely on the manual. This will take a bit longer, but the results will be the same. If the manual does not have adequate examples or if it is too difficult to follow, there are a multitude of books on the market written to help you use your software to its maximum potential.

8.6 EXERCISES

1. To illustrate the simplification afforded by tables, compose a paragraph that places the following tabular information in prose form and discuss the benefits of Table Y.

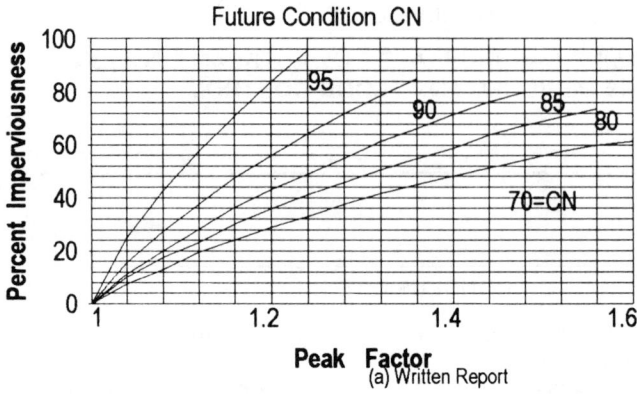

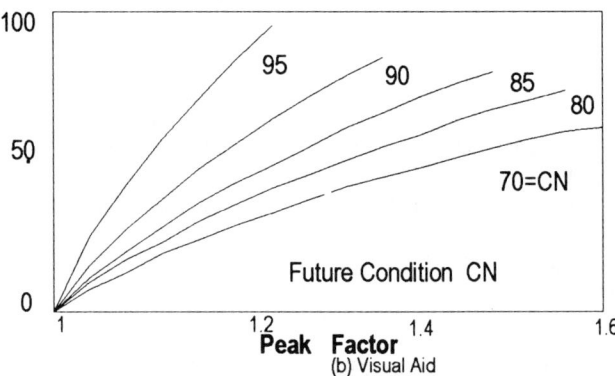

FIG. 34.— *Comparison of Graphical Figure for Written and Oral Presentations*

TABLE Y   Nutritional Content and Cost per Serving of Breakfast Cereals

Cereal	Fiber (grams)	Sugar (mg)	Sodium (mg)	Fat (grams)	Cost (cents)	Calories/ Serving
Corn flakes	T[a]	2	280	0	10	110
Corn bran	5	6	300	1	14	110
Raisin bran	4	9	150	1	13	80
Shredded wheat	3	0	T[a]	1	12	110

[a]T = Trace

2. Find a sketch or drawing in a textbook related to engineering and use it to answer the following questions:

   a. How was the figure introduced to the reader?

```
┌─────────────────────────────────────────┐
│              CONCLUSIONS                 │
│                                          │
│   1. Downstream Erosion                  │
│   2. Storage Volume                      │
│   3. Efficiency                          │
└─────────────────────────────────────────┘
```

FIG. 35.—*Overly Simplified Visual Aid*

    b. Is the information provided by the figure necessary?

    c. Is the information easy to interpret?

    d. Is the title complete and meaningful?

3. Using professional journals from your technical speciality, identify figures and/or tables that have good and poor titles. Discuss your evaluation.

4. In professional journals or textbooks, find two examples of tables that do not communicate enough information to understand the table. Then find two examples of properly structured tables. How do the properly structured tables communicate their information? What changes would you make to the two poorly structured tables to better communicate the information?

5. Summarize the benefits of using tables and figures.

6. Discuss the conditions under which figures are useful for presenting information.

7. Compose a paragraph that discusses the intent of Figure 8-5. Discuss the advantage of presenting the information in the format of a figure.

8. Transform the tabular information in exercise 8-1 to a figure. Discuss the pros and cons of using a figure and a table.

9. Show how the tabular data of Fig. 32 could be presented graphically.

10. Choose an example of quantitative data from your workplace or school. Arrange the data in a table and a bar chart. Which format is more appropriate?

11. Bring in examples of the following visuals from newspapers or magazines: a pie chart, a bar chart, a map, a table. Analyze the visuals for clarity, organization, and effectiveness.

CHAPTER 9 / BUSINESS CORRESPONDENCE

DO

- think through what you wish to say in your memo or letter
- use standard format
- initialize the memo
- date the memo
- type the memo, when practical
- use a pleasant, personable, polite tone
- use "I" and "you" to make a business letter more personal

DON'T

- use novelty papers, inks, typefaces or fonts
- be overly friendly or familiar
- thank the reader in advance for something he or she has not yet done
- use cliches or pompous language
- distribute the memo or send the letter without first proofreading it

---

9.1 INTRODUCTION

Today in any occupation, no matter how scientific or technical, one is expected to write clear, concise, and effective business letters and memoranda. The inability to write effectively is not only detrimental to career advancement, but also expensive for the organization. The personnel, equipment depreciation, and paper required to produce a one-page business letter is close to $9.

Business correspondence is usually of two types: memoranda and letters. Both are written records; without accurate records, an organization would collapse. Every employee should feel responsible for the accuracy of his or her organization's records. This chapter offers some strategies for composing effective memos and letters.

9.2 MEMORANDA

Memoranda are usually brief, one-page messages that are circulated within an organization. Memorandum is the singular form, and memo is a commonly accepted abbreviation. The purpose of memoranda is to create a record in an efficient manner. It can be used to notify an individual, or a group about a meeting or a due date. Memos can also provide written notification of an action to be taken. Even a telephone message is a simple form of a memorandum.

Memos are a popular form of business correspondence for a number of reasons. First, they do not require the formatting expected in more formal business communication. They are also easy to distribute. For example, if you need to notify 10 people of the date, time, and place of a meeting, a memo can be photocopied and placed in each committee member's mailboxes. A memo should clearly document the needs or responsibilities of the receiver. Finally, it provides a written record that the recipient can use for later reference, such as the location of a meeting or directions to an unfamiliar location.

Because it is assumed that memoranda are transactions between individuals in the same organization, it is acceptable to eliminate the return address, inside address, formal salutation, and formal closing, all of which would be necessary in a business letter. Instead, memoranda use a format that indicates the recipient's name and title, the sender's name and title, the date, and a brief description of the subject of the memo. Some organizations use pre-printed memo forms (see Fig. 36), while others allow each employee to create his or her own format. The sender does not usually sign the memo, although he or she should initial it to indicate approval of its content. A memo addressed to an individual may include his or her title following the name; in such cases, the sender's title should also be included.

Most memos are intended to be informal, yet the format is somewhat standardized. The heading material (To, From, Date, Subject) is double spaced, with triple-spacing between it and the body of the memo. Double spacing the heading improves its clarity. The body paragraphs usually follow a block format, which means that the paragraphs are short and single spaced, with double–spacing between paragraphs. The paragraphs are not indented as they sometimes are in more formal letters. An exception to this format is made for very short, one paragraph memos that are double–spaced. For example, if the only intent is to notify the members of a committee about the date, time, and location of their next meeting, then double–spacing would be more appropriate (see Fig. 37).

When creating your own memos use the accepted format shown in Fig. 38. In the nonstandardized format, the date is sometimes placed in the upper right-hand portion of the memo. Note that there is no signature.

Potted Plants, Inc

To: _____

From: _____

Date: _____

Subject: _____

MESSAGE. . . . _____

_____

_____

_____

_____

_____

FIG. 36.—*Preprinted Memorandum*

TO:	Fire Safety Review Committee
FROM:	G. Moglen
DATE:	July 10, 1990
SUBJECT:	Meeting to Review Draft Standards

   The Fire Safety Review Committee will meet on July 31, 1990, at 9 a.m. in the west-wing Conference Room. Please review the attached draft of the Fire Safety Standards prior to the meeting. The Standards will be finalized at the meeting.

FIG. 37.—*One-Paragraph Memo*

Instead, the writer initials his or her typed name. It is important for the writer's to be included in case the memo's recipient has questions. This is especially important in the phone-message memo.

   Memos should cover only one topic at a time. Composing an effective and descriptive subject line is a good way to test one's logical thinking. The subject line should be a summary of the memo. Compare the following three groups of memo subject lines. Which are the more effective?

   1. a. Lunchroom policies.
      b. Restricted areas for employees' food consumption.

MEMO

TO: John West, Coordinator
FROM: Jane Smith, Assistant Director
DATE: July 6, 1990
SUBJECT: Annual Report

Please submit any revenues from your department for inclusion in the annual report. I have enclosed a copy of last year's report to help you in your calculations.

I would like to have your input by July 20, 1990.

Enclosure

cc: Dr. William Jones, Director

FIG. 38.—*Nonstandardized Memorandum*

2. a. Security violations.
   b. Recent burglary in Building 6.
3. a. Required equipment.
   b. A requisition for new microwave transmitting equipment.

In each of these three examples, "a" is too general; the specificity of the "b" examples clearly shows the purpose of the memo.

Read the memo of Fig. 39. What is your impression? Would you be able to take action after reading this? Why not?

Now read the memo in Fig. 40. This is a shorter memo but is it clear? What specific request is being made in this memo? How could this memo be clarified?

9.3   LETTERS

Letters, because they are almost always sent out of the company, are very important. They reflect you and your employer. Poorly written business letters can be detrimental to your company's business volume and to your professional reputation. Letters must be timely, not just accurate and informative. A late letter suggests disrespect or poor time management. An inaccurate letter may lead to incorrect decisions, possibly even legal action against you or your employer. Even typographical errors may lead to problems, such as a letter offering a worker $52 per hour, rather than the intended hourly wage of $25. In some cases, the typographical error might be legally binding.

Department of Occupational Services

Date: Apr 7, 1988
From: Director, Office of Management
Subject: Delegation of Authority to Determine Action on Within-Grade Increases
and Delegation of Authority to Determine Action on Quality Step Increases

To: PHS Agency Heads

Authority Delegated To Whom

Pursuant to the authority delegated by the Assistant Secretary for Health to the Director, Office of Management, PHS, on Mar 28, 1988, I hereby delegate to the PHS Agency Heads, the authority to redelegate to some official other than the immediate supervisor the authority to reverse or sustain a negative determination upon request of an employee for reconsideration of a negative determination for periodic within-grade increases for personnel within your respective organizations.

Redelegation and Restrictions
1. The authority delegated above may not be redelegated.
2. On December 2, 1986, ASPER delegated directly to all immediate supervisors:
   (1) the authority to make favorable and negative level of competence determinations for periodic within-grade increases for the employees under their supervision; and (2) the authority to recommend quality step increases and to concur and nonconcur in quality step increases for the employees under their supervision.

Information and Guidance
Information and Guidance for exercising this authority is found in 5 CFR 531, Subpart D, and HHS Instruction 531-4.

Prior Delegations
This delegation supersedes the November 18, 1982, Delegation of Authority to Determine Action on Within-Grade Increases from the Director, Office of Management, PHS, to the PHS Agency Heads and the November 30, 1982, Delegation of Authority to Determine Action on Quality Step Increases from the Director, Office of Management, PHS, to the PHS Agency Heads. To the extent that previous delegations and redelegations made to other PHS officials of the authority to determine action on within-grade increases and quality step increases are consistent with the provisions of this delegation, they may remain in defect for no longer than 90 days from the effective date of this delegation.
Now, practice rewriting the following memo for organization and emphasis.

FIG. 39.—*Poorly Written Memo*

MEMO

TO: Company Personnel
FROM: C. Smith
SUBJECT: Meeting On Strategic Planning Skills
DATE: October 6, 1990

On November 7, the Wilson Corporation will host a seminar on Strategic Planning Skills. Approximately 200 participants are expected to arrive at the North Building Auditorium. This may cause a bottleneck at the Lake Street intersection. Please enter at the East Gate at the Glendale Street intersection. Also, the cafeteria in the North Building will be crowded at lunchtime.

All employees are urged to take effective steps to prevent these problems.

FIG. 40.—*Memo Requesting Action*

9.3.1 Purpose of Letters

Professionals write business letters for a variety of reasons. Most often, a letter must address a specific topic and a specific person; thus, a form letter may not suffice. Also, the wide variety of reasons for writing letters means that every letter will require individual attention. The following are a few of the types of letters that professionals must write:

1. Letter of introduction: Its purpose is to introduce a person, product, or service. Such letters are common in sales and promotional efforts.
2. Letter of transmittal: A letter of transmittal is used when forwarding project reports, minutes of business meetings, etc. In some cases, a memo is used in place of the more formal letter; for example, when the information being transmitted is in a standardized form, such as meeting minutes.
3. Letter of inquiry: A letter of inquiry is used to obtain information about a person, product, or service, for example, the cost of a product, the availability of a computer service, or the interest of a person who has a particular expertise.
4. Letter of progress: When a company deals with a client or is a subcontractor, regular progress reports may be required. Where the extent of notification is minimal, such that a complete report is not necessary, a letter of progress can be used.
5. Letter of recommendation: A letter of recommendation is used to recommend a person or a course of action, that is, a decision.
6. Letter of approval: For some decisions, outside approval is required. A letter of approval for decisions, draft reports, meet-

ing times, etc., may be necessary before certain business activities can continue. In some cases, the letter of approval may represent a legal authorization to proceed with some activity.

Although most people learned the six major parts of a business letter in elementary school, many educated people are still unsure about the exact format. The six parts of a business letter are: the heading (the writer's address and the date), the inside address (the recipient of the letter), the salutation (Dear Sir or Madam, or another appropriate greeting), the body (the content), the complimentary closing, and the signature.

When a letter without either a heading or an appropriate signature is sent, a negative message is also sent. All working professionals should communicate respect to the recipient of their letters by structuring those letters in the commonly accepted style. Be sure to do the following:

1. Select white bond paper of about 20-lb. weight. Do not use flimsy paper, unless you are doing business internationally and are using airmail stationery.
2. Choose conventional typeface. Avoid italics, block letter, or desktop publishing lettering for formal correspondence. Avoid *any* desktop graphics.
3. Use single-spaced letters. Use either full-block format (Fig. 38) or modified block format (Fig. 39). Full–block format is the easiest for a writer unsure about letter format. Text is flush left, with double–spacing between paragraphs.
4. Modified block is more sophisticated (see Fig. 39). The heading and the complimentary closing begin in the middle of the page and line up with each other. Paragraphs are indented five spaces and the letter is single-spaced.
5. Occasionally a subject line is used. Unless this is a required company style, omit a subject line and use the first sentence to state the subject of the letter.
6. Add initials after the signature line only if someone else types the letter. Then the writer's initials are capitalized, followed by a colon and the typist's initials.
7. If other people are receiving copies, add "c" or "cc", a colon, and that person's name. "cc" stands for carbon copy, so technically a single "c" is sufficient.

There are, of course, innumerable variations. All employees should learn the style their company prefers and use that style in official correspondence.

Do not be hesitant about writing several drafts. Remember, it is easier to write a long, rambling letter than a short, well organized one! But the well organized letter will be more effective and make a more positive impression on the recipient.

In general, the body of a letter should contain three parts:

1. A brief introduction or opening that clearly states the purpose of the letter;
2. Several paragraphs containing complete sentences and specific details; and
3. A polite closing that assists the reader to take some action steps.

9.4   OPENINGS AND CLOSINGS OF BUSINESS CORRESPONDENCE

Openings and closings of letters, memos, and reports deserve special discussion because they are in positions of emphasis, i.e., they create the first and last impressions the reader gets from the piece of writing. Unfortunately, writers who wish to avoid cliche beginnings and endings often have difficulty starting and finishing their memos and letters.

9.4.1   Beginnings

The opening sentence should educate and entertain by: (1) Identifying the subject; and (2) catching the reader's interest. Most readers of business writing are preoccupied with other business when they begin to read a memo, letter, or report. Therefore, you must catch their interest and focus their attention on your subject. Even if the letter's recipient is required to read what you've written, catching his or her interest immediately will help ensure that close attention is paid. To persuade the reader, you must also catch his or her attention.

Ideally, the first paragraph should perform these three functions:

1. Get favorable attention;
2. Indicate what the purpose is; and
3. Link up with previous correspondence by referring to a date or subject.

If the first paragraph is direct and interesting, the reader is likely to read the entire letter. If it is not, the rest may be skimmed or skipped entirely. To be effective, the first paragraph should be short and substantive.

Never put more than two or three sentences in your first paragraph. The reference to the date of earlier letters should always be subordinated. A surprising number of writers begin letters with sentences such as:

> This is to answer your letter of October 10.
> We have received your letter of October 10.
> Referring to your letter of October 10.

Why waste the most important part of a letter—the equivalent of a newspaper headline—merely to tell the reader that the letter was received, or that it was dated October 10? The important task of the first paragraph is to announce the subject of the letter. All else should be subordinate. Note the effectiveness of the second method of writing in each of the following opening paragraphs:

Opening	Comments
Replying to your letter of May 10, we can say that our research staff has been working on the problem you mentioned, and we have resolved it.	Weak and ineffective, because the first 10 words tell the reader nothing new
Our research staff has solved the problem of insulating old homes, about which you inquired in your letter of May 10.	Direct and effective
Acknowledging receipt of your letter of May 15, in which you asked for a copy of our regulations.	Incomplete sentence
We are enclosing a copy of our regulations, which you requested in your letter of May 15.	Complete
Your letter of January 2 was received and reviewed. We are referring your question to our sales department.	Too blunt and short; unpleasant tone; suggests a lack of interest in their concerns
Our sales department is assembling material that should be helpful in answering your inquiry of January 2.	More positive tone

The following sentences are good beginnings:

Thank you for your request for information about our hiring policies.

The catalogue you requested on May 27 was mailed today.

Here is the bulletin you requested.

The material on page 16 of the enclosed brochure will answer your question in your letter of June 4.

As soon as we received your letter, we wired our New York office to forward the information you requested.

9.4.2 Endings

The closing paragraph not only ties your writing together and ends it emphatically, but it may also make a significant point. A closing may recommend a course of action, offer a value judgment, speculate on the implications of your ideas, make a prediction, or simply summarize your main points.

One principle governs the writing of the final paragraph: Stop when the message is complete. A good closing is concise and ends the letter emphatically. Don't close with a cliche. Also, don't introduce a new topic in your closing. A closing should always reinforce the ideas presented in the other paragraphs of the letter.

The most ineffective of all closings is the "-ing" ending. It is weak and incomplete. "Thanking you in advance" and "Trusting to have your cooperation in the matter" are examples of this type of weak ending.

Instead of the "-ing" ending, try using an ending similar to one of these:

1. If you mail us your check today, we will ship your order on Thursday.
2. Please sign your name at the bottom of this letter and return it in the enclosed envelope.
3. Will you let us know by April 14, so that we can place your order?
4. All of us wish you success in your new job.
5. We appreciate your cooperation.
6. Thank you for being so prompt in getting this information to us.

9.5   THE LETTER WRITER'S GUIDELIST

The following 16 questions provide a checklist for proofreading your correspondence. A "no" answer to any of these questions indicates a trouble spot that needs attention.

1. Are most of your letters less than a page long?
2. Is your average sentence less than 22 words?
3. Did you try to keep paragraphs short—less than 10 lines?
4. Did you avoid beginning a letter with "Reference is made" or "This office is in receipt of your letter"?
5. Do you know some good techniques for beginning letters naturally and conversationally?
6. Can you think of four words that will take the place of the word "however"?
7. Do you paraphase laws and regulations instead of playing safe and quoting them?

8. Do you use personal pronouns freely, particularly the personal pronoun you?
9. Are your letters written in the first person (we (I) shall appreciate) rather than the third person (this Bureau will appreciate)?
10. Do you use active verbs (the manager read the letter) rather than passive ones (the letter was read by the manager)?
11. When you have a choice, do you choose little words (pay, help, mistake) rather than big ones (remuneration, assistance, inadvertent)?
12. Whenever possible do you refer to people by name (Mr. Jones, Ms. Smith) rather than categorically (the claimant, the veteran, the applicant)?
13. Does your letter sound as you do when you speak in a careful manner?
14. Do you answer a question before explaining the answer?

9.6   EXERCISES

1. Arrange the material below into a suitable and readable memorandum. Omit all unnecessary words.

   A reminder from you to all men and women who are bank managers. Three more meetings will be held because discussion indicated there was interest in these areas. The three meetings are all a continuation of the Manager-Customer Conference. We will have three speakers who are really outstanding. Professor Harry Baker of the University of Michigan is going to talk on "The First and the Last Loan." That talk will come on March 27. Mr. Potter Reid who works for Reid, Shaw, and Sloan, Inc. will talk on April 24 on the topic, "The Bank's Customer and the Bank's Language." On April 1 Ms. Marilyn Jones speaks on "A New Era: The Bank Card." Ms. Jones is a Bank Americard executive. Conference Room C will be the site of all of these meetings. The time will be 9:30 a.m. If you can come, don't send any notice; if you cannot come, let me know before March 20.

2. At least three people have misused (or abused) sick leave and you have been asked to write a memorandum to all people in your department stating the policy on sick leave. Be simple, clear, concise. Remember, the object of the memorandum is to stop the misuse of sick leave.

3. Your supervisor told you to analyze copies of all correspondence sent out by your division (or department) during one week. You did so and found four major faults: wordiness, misspelling, poor

sentences, weak organization. Write a memo to your supervisor with your conclusions and recommendations.

4. Rewrite the memorandum of Fig. 39 so it is concise and easy to read but accurately presents the necessary information.

5. Write a memorandum for one of the following:
   a. A memo of transmittal for the draft of a sales report. The memo will be distributed to all sales managers in the company. Request (1) comments by specific date; and (2) recommendations based on the sales report about increasing the market share of a product with poor sales.
   b. A memo of transmittal of a first draft of a brochure announcing an upcoming short course to be sponsored by your company. You want the design engineers who will participate in the course to check all information in the draft, provide additional details of the content of their presentations, and indicate the approximate length of their presentation, including how they will use the time (lecture, problem solving, case study, review, etc.).

6. Compose memos for the following situations:
   a. Your department has submitted to you an unreasonable requisition for new equipment. List the items of which you do not approve and explain your reasons.
   b. The conditions at your workplace are unsafe. Cite three specific unsafe conditions and briefly suggest a solution that is relatively inexpensive.

7. What is your reaction to the following openings? Which ones are good, fair or poor?
   a. You made several errors on the reply form you submitted, and you neglected to include the signed receipt.
   b. We are in receipt of your memo regarding the four 123 wheels which you say are defective.
   c. Referring to yours of August 9th.
   d. As requested in your letter of October 1, we are enclosing three copies of our latest administrative directory.
   e. This is in reply to your letter of September 15, requesting us to review the application which was previously declined.
   f. We have received your letter of October 9.
   g. Thank you for your helpful suggestions about our upcoming conference.

8. Compose the first paragraph of the letters to address the following situations.
   a. An answer to a letter dated August 29 requesting a copy of a catalog of the courses offered and an application blank.
      The catalog and application blank are being sent today.

b. An inquiry as to the status of the information on employee absenteeism that you requested four weeks ago. The information is needed for a longer report which you are preparing and which is due next week.

c. A reply to an applicant for a position in your department. The applicant was well qualified, but another person was hired for the position.

d. A request for a fellow worker to serve as chairman of the annual blood bank drive. The fellow worker has indicated that s/he would be willing to take on this task.

e. An acknowledgement of a request for information. You are unable to provide the information, but think that the Department of Human Resources can do so.

f. A thank you to a local businessman for agreeing to serve on a panel of experts for a meeting you are arranging.

g. An announcement of the retirement of an employee who has been with the organization for 43 years. His name is Harry Miller and he works in the Department. He is well known and well liked.

9. Compose the last paragraph of the letters for the following situations.

a. You are applying for a position with a large government contractor and are sending your resume. You would like an interview as soon as possible.

b. You are recommending that your department order new filing cabinets which are costly but which will last for many years and which are necessary to hold the new forms your department is using.

c. You are requesting information on several cost-saving devices. You need the information for a conference you will be attending next week.

d. You are replying to a complaint you received from a displeased congressman. The congressman's office is receiving hundreds of letters from angry constituents who believe the relocation of a shipbuilding facility will cause the loss of their jobs.

e. You are making reservations for your supervisor at a hotel in New York. Your supervisor will be attending a meeting in New York next week, but he will not be in the office again before going to the meeting.

f. You must decline a request to serve as chairman of the employee bowling league. You have taken on additional job-related duties. You do know someone who is willing to assume the duties of chairmanship.

g. You are requesting fellow employees to participate in a collection of food and clothing for those who are less fortunate.

The drive will take place during November. You need to let the employees know where to bring their donations.

h. You are answering a request for information. The request was not very specific and you have tried to answer all of the questions and give as much information as you could.

i. You must apologize to a local businessman who visited your office last week and was splattered with paint by the careless painters who were working in your office. The businessman did not realize that his suit had been splattered until later on that evening, and by then it was too late to remove the paint.

10. Write a memo to the instructor of one of your courses advising the instructor on the sources of information you will use for an assigned project.

11. Obtain a memo from a professor or from someone in your departmental office. Present a critique of the effectiveness of the memo. How could you rewrite the memo to be more effective?

CHAPTER 10 / **RESUMES**

DO

- underline, capitalize, or italicize where emphasizing is desired
- use action words in describing employment, leadership positions, and skills, such as: analyzed, directed, evaluated, prepared, supervised
- use phrases rather than whole sentences
- be honest
- send a cover letter with the resume

DON'T

- use the word "I"
- use script type
- use lengthy descriptions or complete sentences
- provide extraneous information such as the word "RESUME" at the top of the document
- include personal information
- use gimmicks such as brightly colored paper to catch the employer's attention
- be modest

10.1 INTRODUCTION

A resume is a document that summarizes your educational and work experience to provide a prospective employer with information on what you have to offer the company. The main purpose of the resume is to convince the prospective employer that you should be considered for the position. The employer will initially scan your resume, typically for 30 to 60 seconds, and place it in either the "consider" pile or the "reject" pile, depending on whether the resume shows that you will be one of the best candidates for the job. Therefore, your resume needs to be the best in the pile! The resume must be neat, well organized, and concise so that in those first seconds the employer can determine whether or not your education and experience will meet the company's needs.

Before beginning a resume, gather all information on your educational background and work experience. Next, organize this information and write the resume. Then, type and check the final product. Finally, write a cover letter to accompany your resume. Each of these steps is discussed in more detail in the following sections.

10.2   STEP 1: GATHER INFORMATION

The main components of the resume are your name and address, educational background, work experience, and awards and honors. References will be required, but for an entry-level position it is not necessary to list your references on the resume. The employer will usually not contact references until he or she has spoken with you.

Assistance in developing a complete pre-employment file is often available through your university or college career and placement office. For a modest fee, they will make up a packet of reference letters and transcripts that can be forwarded to a prospective employer. By using the placement center, you need only ask an individual for one reference letter; this is placed in your file and sent out until you replace it with other letters.

Personal information should not be included on the resume. This is not information that the employer will need in deciding who is the best candidate for the position. In addition, personal data may be discriminatory information that the employer is not legally permitted to request. You should not include information about your age, your ethnic group, your health, or your marital or family status. Also, do not include a photograph.

Gather the following information before beginning to write your resume:

10.2.1   Name and Address
1. Formal name, including middle name or initial (avoid using nicknames)
2. Address and phone number where you can be reached. You may want to include two addresses and phone numbers if one of those is a temporary address, such as a school address. If your address changes, you should revise your resume and send a copy of the new resume to those for whom you might want to work.

10.2.2   Educational Background
1. Colleges and universities attended
2. Dates of attendance
3. Degree obtained and date (or the date when the degree is expected to be conferred)
4. Major subject (minor subjects relevant to the employer's needs)
5. Grade point average, both overall and major subject

6. Awards or honors acquired during your college education (do not go back as far as high school). If you have more than one or two awards or honors, include them in a separate section.
7. Special projects completed, such as building a concrete canoe
8. Leadership positions held (e.g., officer in honor society) if you held only one or two positions. Otherwise, include it in a separate section with awards and honors.

10.2.3   Work Experience

1. Descriptive title of any positions held, including summer work
2. Name of firm or agency of employment
3. Dates of employment and any promotions
4. Primary responsibilities
5. Contributions made to the firm
6. Skills developed on the job (e.g., knowledge of computer software)
7. Supervisory responsibilities, if applicable

10.2.4   Awards and Honors

1. Academic honors (do not include high school) including titles, dates and reason for the award (service, scholarly achievement, etc.)
2. Professional awards
3. Professional society membership, including student chapters. You should also include your position in the organization.

10.2.5   References

1. Three to five names, addresses, and phone numbers of people who will provide references for you. It is important to ask these people for permission before using their names as references. It is also advisable to provide your references with a copy of your resume so that they will have a better understanding of your experience and education.

10.3   STEP 2: WRITING THE RESUME

Before writing your resume, spend some time thinking about the appearance and organization of this important document. What do you wish to stress to the employer: education or employment? A recent college graduate should usually stress education since his or her employment record is usually sparse.

For an entry-level position, the employer will probably skim over your resume in less than 60 seconds and decide whether or not to consider you for an interview. Therefore, your resume needs to be reader-friendly. It has to be well organized and placed in a format that is easy to read.

The use of capital letters and underlining for headings makes the sections of the resume more noticeable, thus making the resume easier to follow. Fig. 41 is an example of a well organized resume. Each section stands out because of the headings, making the resume easy to skim.

A resume for an entry-level position should generally be kept to one page. It may be longer under certain circumstances such as when it was requested or where only a handful of applicants are submitting resumes. In these cases, the employer will spend more time looking at your resume.

A basic resume will have six major sections: name and address, objective, education, work experience, honors and awards, and references. Each of these is discussed in more detail in the following sections.

SUSAN B. PARKER

Local Address: *Permanent Address:*

Smith Hall, Room 82   452 N. Shore Road
University of Maryland   Annapolis, MD   22043
College Park, MD   20742   (301) 322-9898
(301) 454-1021

OBJECTIVE   Entry-level civil engineering position

EDUCATION   Bachelor of Science
                                        Major: Civil Engineering
                                        University of Maryland
                                        Expected date of graduation: May, 1991
                                        GPA: 3.35

EXPERIENCE   Computer Programming Assistant, University of
                                        Maryland, 5/90–8/90. Aided in development of computer program to analyze transportation routes, entered data for routes, and assisted in analysis of output.

                                        Bookstore Salesperson, University Bookstore, College Park, MD, 5/89–8/89. Handled sales of technical books.

AWARDS AND HONORS Vice-president, ASCE student chapter, 9/89–5/90
                                        Member, Chi Epsilon
                                        ASCE Scholarship (Senior year)

REFERENCES   Available from University Placement Office

FIG. 41—*Example of a student's resume.*

10.3.1 Name and Address

The first piece of information that the reader should see is your formal name in capital letters centered at the top of the page (refer to Fig. 41 for the format). Your name should be prominently displayed so a prospective employer will not confuse you with someone else. Below your name, your address and phone number should be given, along with an alternative address and number if the first is a temporary one. The address must be a complete, including the zip code, in case a prospective employer sends something to you through the mail. If you have changed your name during your academic or professional career, you may either state this information directly below you name on the resume (e.g., Formerly, Susan B. Jones), but only if you feel that this information is pertinent to the resume. Otherwise, when transcripts or other documentation bearing your previous name are requested, you should include a letter that states your previous name.

10.3.2 Objective

A statement of your objective is next on the resume. The purpose of listing your employment objective is to provide a prospective employer with a statement of the type of employment you are seeking. This makes it possible to match your needs with those of the company. For example, if the purpose of sending the resume is to get an interview for an entry level position, the objective may read:

OBJECTIVE: Permanent position as a mechanical engineer in automotive design.

If your objective is to obtain a summer position, that should be clearly stated instead. For example:

OBJECTIVE: Temporary position as a computer programmer.

10.3.3 Education

Most recent graduates will stress education rather than work experience. However, if you feel that your work experience is more important than your educational background, you may place work experience next on the resume, followed by a summary of your education. Fig. 41 is an example showing the educational experience before the work experience.

Colleges or universities attended, particularly those where a degree was awarded, are listed in reverse chronological order (i.e., the most recent first). The most important piece of educational information is the degree obtained, so place this first along with the academic major and the date of graduation (or expected date).

Following the degree, identify the name of the university or college, the city and state where it is located, and both your overall and major subject grade point average as shown in Fig. 41. If your grade point average is low, do not include it on the resume since it is not a positive selling point. You may also want to list one or two courses that you believe demonstrate some particular knowledge or skill.

There are many types of educational experiences possible. If you hold more than one degree, list the most recent first or in the case of a double major, list first the degree which is more relevant to the position for which you are applying. The following examples show how to list two degrees obtained at different times. In the first example, two different bachelor degrees were obtained:

EDUCATION: Bachelor of Science
Major: Electrical Engineering
Pennsylvania State University
Expected date of graduation: May, 1992
GPA: 3.40

Bachelor of Arts
Major: History
Arizona State University
Date of graduation: May, 1985
GPA: 3.22

In the second example, both a bachelor and a master degree were obtained:

EDUCATION: Master of Science
Major: Physics
Colorado State University
Expected date of graduation: Dec., 1992

Bachelor of Science
Major: Physics
Colorado State University
Date of graduation: May, 1990

The following example shows how to list two degrees obtained at the same time:

EDUCATION: Bachelor of Science
Major: Geology
New Mexico State University
Expected date of graduation: May, 1992
GPA: 3.80

Bachelor of Arts
Major: Russian
New Mexico State University
Expected date of graduation: May, 1992
GPA: 3.65

You may also want to mention honors, academic memberships, and leadership positions held at this university. However, if there are more than one or two, they should be placed under the heading "Awards and Honors" later in the resume.

10.3.4 Work Experience

For most college graduates, this section will be rather skimpy. You should include both relevant and nonrelevent experience. Even though working in a bookstore during the summer may not be relevant to a position as an aerospace engineer, an employer likes to see that you have worked outside the university. It indicates that you have held a position with some responsibility. Someone anticipating graduation from college may want to include positions held during high school, but these should be included only if they indicate a position of responsibility. A job as either a paperboy or a babysitter would not be appropriate.

Work experience is listed in reverse chronological order, as shown in Fig. 41. The first piece of information is a descriptive title of your position, followed by the name of the firm or employer, the city and state, and the dates of employment. There is no reason to provide a reason for leaving, although this is something you should think about before going on to interview. The interviewer may ask this question if the termination date does not coincide with a recognizable date such as the start of a new academic year.

A brief description of your work while employed by each firm is next. The description should include only phrases rather than complete sentences. Be sure to use action words such as those listed in the DO section. Lists of responsibilities should be presented in a balanced form. For example, consider the following:

> *XYZ Corporation, Apprentice Programmer*
> Compiled data bases; debugging programs; was responsible for maintaining the software library.

This is unbalanced because the first two entries differ in tense and the last entry is in more of a sentence form. The following is balanced, as well as less wordy:

> *XYZ Corporation, Apprentice Programmer*
> Compiled data bases; debugged programs; maintained software library.

Active, rather than passive constructions should be used. For example, consider how the following passive phrase has been rewritten to the active construction.

> ***** WRONG *****
> was responsible for conducting laboratory experiments
> ***** RIGHT *****
> conducted laboratory experiments

It is important to keep the descriptions very brief; the employer does not have the time or the interest to read lengthy job descriptions.

10.3.5 References

Listing references at the end of the resume is optional for entry-level positions. The employer will usually not contact references until after your interview, so the references may be provided later. Resumes directed at advanced positions should include three references.

Whether or not the references are listed, be sure to have the permission of at least three people to use as references before you send out the resume. Supply each of the people providing references with a copy of your resume so that they will be prepared when an employer calls. Your references should know your entire background so that they can speak with authority about your abilities.

10.4   STEP 3: THE FINAL PRODUCT

Now that your resume is written, have your advisor or a friend read it over to make suggestions and corrections. As you carefully reread the resume, check for the following:

- typographical errors
- grammatical errors
- correct name and phone number
- neatness and organization

Your resume should be either typed using a new black ribbon or, if a word processor is used, printed using a high quality printer such as a laser printer. The resume should be copied onto good quality white or off-white paper. Remember—NO COLORED PAPER!!!

Before sending the resume, take some time to write a cover letter, as described in the section below, both to introduce yourself and to sell the employer on the idea that you should be considered for a position. The resume and cover letter should be mailed in a 9 x 12-inch manila envelope to avoid creasing.

Be sure to keep a file on your job search. The file should include:

- copies of your resume (all versions)
- copies of cover letters (these letters will contain information on the companies to whom you sent resumes and dates of mailings)
- dates of follow-up letters
- dates of all phone conversations with employers, as well as the name of the individual you talked with and the major points of the discussion
- letters of reply from potential employers

Refer to the list of do's and don't's at the beginning of this chapter as you write your resume.

10.5   STEP 4: THE COVER LETTER

The cover letter is a very important part of your employment package. Every resume should be accompanied by a letter to introduce you and to sell you as a well qualified potential employee to the employment recruiter so that he or she will be interested in reviewing the enclosed resume. A well written cover letter greatly increases your chance of getting an interview.

The cover letter may be organized similar to the one in Fig. 42. No more than four short paragraphs should be used. The paragraphs may contain the following information:

Paragraph 1: Introduce yourself, including your degree (or anticipated degree), when and where that degree was obtained, and the objective of sending your resume (what position is desired, etc.). Also state when you will be available. If you are applying for a summer position, specify the ending date.

Paragraph 2: Describe your motivation for writing the letter. For example, you might have been motivated to send the resume because of a suggestion by one of your professors, because of a person or display at a job fair, or because someone from the company spoke at a meeting.

Paragraph 3: Emphasize your special qualifications relevant to the job of interest. These could include course work, special skills obtained through work experience, or a skill acquired through working on a project at school. You may also include your grade–point average, if you believe it will encourage the reader to take a closer look at your resume.

Paragraph 4: State your interest in the company or agency. Inform the reader that you will follow up with a telephone call in several weeks. You may also include your telephone number. Even though your telephone number appears on the resume, repeat it so that the employer will not have to look for it if the letter and resume get separated.

Now a few tips for a well written cover letter:

1. Use business letter format as shown in Fig. 42 (also see Chapter 9).
2. Type the letter (handwritten letters are unacceptable) and check for typographical errors. Never send a photocopy of a cover letter.
3. Although the resume will likely be the same for each company you send it to, the cover letter should be personalized for the particular firm.
4. Make sure the letter is short, not exceeding one page. The employer will not be interested in reading a lengthy letter.

April 14, 1991

452 N. Shore Road
Annapolis, MD 22043

Mr. John G. Graper
Personnel Director
AAA Engineering Consultants
Silver Springs, MD

Dear Mr. Graper:

I am interested in an entry-level position with AAA Engineering Consultants. I am currently a student at the University of Maryland in the Civil Engineering Department. I expect to complete my Bachelor of Science degree in May, 1991, and will be available to assume a position June 1, 1991.

I am sending my resume at the recommendation of my advisor, Dr. James Smith. He informed me that there are several openings with your firm.

My course work and summer employment will be valuable to me in solving problems as an employee of AAA Engineering Consultants. I have maintained a grade point average of 3.35 in my overall course work and 3.75 in my engineering course work. During this past year, I have been the vice-president of the student chapter of ASCE. During the summer of 1990, I developed computer skills, particularly in the Fortran and C languages.

I hope that my background is of interest to you and that I will hear from you in the near future. I may be reached at (301) 454-1021 (school) or (301) 322-9898 (home). If I have not heard from you within several weeks, I will call you. I am looking forward to hearing from you.

Sincerely,

Susan B. Parker

FIG. 42.—*Cover letter for a student's resume.*

5. Always get the name and title of the appropriate person, even if the individual's name does not appear in the job advertisement. Never write "To whom it may concern." If absolutely necessary, use "Dear Sir or Madam."
6. If someone has recommended you, emphasize that person's name.
7. Be positive. Never bring up weaknesses in order to excuse them.

You're all set to send the resume and cover letter. Now sit back and wait for the telephone to ring!

Bill Williams
9 Duffryn Avenue
Malvern, PA
Phone: 123-4567

Education: B.S. Civil Engineering, Wright College, May, 1989
GPA: 2.16

Experience: Short-order cook, Twaddell's Diner, Ocean City, MD
Summer, 1984

Construction Inspector, Dept. of Highways, Casper, WY,
Summer, 1985
Responsible for soils-testing, pile-driving inspection, concrete-strength testing

Draftsperson, XYZ and Associates, Van Nuys, CA,
Summer, 1986, Summer, 1987
General drafting for land developing plans

Computer Programmer, XYZ and Associates, Van Nuys, CA,
Summer, 1988
Fortrans and Basic programming; developed data base management system

Personal Data: 23-years old; hobbies: sports, mountain climbing; height 5ft 9in., weight, 150 lbs,; good health

Reference: Professor John Doe
Dept. of Civil Engineering
Wright College
Wright, CA

FIG. 43.—*Example of Young Professional's Resume.*

10.6   EXERCISES

1. Visit your college or university placement office. Obtain copies of three sample resumes. Critically evaluate their content and format. Discuss the positive and negative aspects of each resume.
2. Using the resume of Fig. 41, place the information into a letter that would be used instead of the resume. Discuss why the resume is a better instrument than the letter.
3. How could the resume of Fig. 41 be modified to indicate that Ms. Parker has knowledge of several computer languages (e.g., FORTRAN, BASIC, C) and several software packages (e.g., Lotus, dBase III, WordPerfect)?
4. Write a special cover letter that would accompany your resume for a position as a summer intern with a state legislator.
5. Having sent your resume with a cover letter to a prospective employer, write a letter to accompany the list of references specifically requested by the personnel director of the company.
6. Having received a written offer for a position, write two letters: one accepting the position and one rejecting the position.
7. A friend provides you with a copy of his resume, which is shown in Fig. 43. Evaluate the resume and recommend changes. Provide reasons for all of your recommendations.
8. Gather all the information listed in Section 10-1. Place this in a file and add to it as you acquire more experiences. This way, you will be ready to write your resume when the time comes.
9. Write a cover letter to a fictitious company using Fig. 42 as a guide. Keep the letter in your file along with your information for your resume.
10. Imagine yourself at age 50. Compile a resume of your accomplishments.
11. Find an engineering position advertised in a newspaper or trade journal. Write a cover letter to the firm requesting a summer job.

CHAPTER 11 / JOB INTERVIEWS

DO

- arrive 5 minutes early
- be enthusiastic, positive, and energetic
- make eye contact
- know the interviewer's name
- let the interviewer lead
- listen to what the interviewer has to say
- ask questions
- take pertinent notes
- bring copies of your resume
- obtain a business card from the interviewer
- dress appropriately
- go to the interview alone

DON'T

- smoke
- chew gum
- jingle pocket change
- act nervous
- speak negatively about school or prior employers
- focus on the salary
- slouch, look bored or distracted

11.1 INTRODUCTION

It is not uncommon for today's college student to take on part-time work during college. Success at finding a good part-time job will most likely require an interview. Since most college students do not have a significant past record of employment, the job interview takes on special importance. But the search for full-time employment as graduation nears is far more important. Again, the interview is a major factor in successfully getting a job. For either part-time or full-time work, you should prepare a resume and be ready to interview with recruiters on campus or elsewhere. It is very beneficial to take some time before going to the interview to prepare properly. This will be helpful in two principal ways: (1) If you have done your homework, the interviewer will be favorably impressed; and (2) properly prepared, you will be more confident and more relaxed.

11.2 PREPARATION

In order to properly prepare yourself to project a self-confident image during a job interview, you need to:

- formulate answers to commonly asked questions
- develop a list of questions to ask the interviewer
- obtain information about the company
- make a critical assessment of your appearance

11.2.1 Interview Questions

Many questions are commonly asked in an entry-level job interview. Have a friend or your advisor ask you some of the questions provided below, so that you will have some actual practice at providing verbal responses.

The best answer to any question is a positive one. Think about how you can turn a question around so that a positive answer is possible. For example, the interviewer may ask you what you think your greatest weaknesses are. Either of the following responses may be considered by an interviewer as positive: (1) You tend to work and think about little else until a project is finished; or (2) you find it difficult to overlook details, no matter how small they are. Giving positive answers helps maintain a favorable interview and prevents the interviewer from dwelling on negative aspects of your background or work habits.

Some questions will be straightforward and relatively easy to answer. However, other questions may be quite difficult. If the interviewer asks you a difficult question, stay calm and give a positive, short answer. The more information you give, the more follow-up questions the interviewer can ask on that subject.

A list of commonly asked interview questions and response suggestions are given below. You should begin to think about responses to these questions at least a week prior to the interview; this will give you adequate time to formulate good responses and recall appropriate examples from your past experiences. Not all of these questions will be asked and the interviewer will undoubtedly have some that are not on the list, but this exercise will help you practice answering questions in a positive, concise way.

1. Tell me about yourself. The interviewer is not interested in personal information. He or she is looking for your strengths and

special skills, your past accomplishments, and your goals for the future.

2. Why are you interested in this job? Discuss how your special interests, knowledge, and skills will be useful in this job and how the type of work is what you are looking for. In preparing for the question, try to identify the types of positions available (e.g., research, design, computer modeling, management, etc.). Then identify facets of each type that you would have positive feelings about.

3. What salary level are you seeking? During the initial interview, you want to minimize your response on this subject. You might give a vague answer such as mid 20's, or upper 20's. But you should not commit yourself to an upper or lower bound.

4. What are your job priorities? Describe what is most important to you. Is it moving into a management position or developing some particular skills? Orient your response toward the specific job, but be honest.

5. Have you had any leadership experience? Describe student activities or organizations you have led or had a part in leading. If you have led an activity outside of academics, mention this also to demonstrate your versatility. If you have not held a formal position of leadership, you might mention your organizational responsibilities on a group project in one of your classes. This is preferable to a response indicating that you have no leadership experience.

6. Have you ever done anything especially creative? Recall any class project that required creative thought. If you have creatively solved some sort of problem, mention it. It does not have to be a technical or academic problem or project.

7. Are you planning to go to graduate school, either part-time or full-time, after working for a year or two? If you have any thoughts of going to graduate school in the near future, discuss this. Most employers will encourage further education.

8. What can you do for us? The interviewer wants to know how you can contribute to the company. You will contribute by bringing any special skills or knowledge you have, as well as enthusiasm and hard work to the job, and by finding ways to use your knowledge to help solve problems.

9. How do you deal with pressure? Briefly mention how you exercise (jog, swim, etc.) or breathe deeply or count to 100. Don't mention consuming alcohol or smoking.

10. How will you stay abreast of current advances in your technical specialty? Mention the names of a few professional journals you

read. It may help you appear more professionally mature. If you
don't read professional journals, do so prior to the job interview.

11. How does your grade point average reflect your abilities? If you
have a high GPA, you might discuss a few items that you are
not strong in, but hope to work on as part of your job. If your
GPA is lower, you might want to discuss how you have matured
during your last year or so of college.

12. What are your strengths/weaknesses? Strengths might be characteristics such as hard working and patient or they might be
special skills that you have acquired in college, such as computer
abilities. Weaknesses should, to the extent possible, be made
positive, as described earlier.

13. Why should this company hire you? Your answer should include
the word "contribute." You want to contribute and be a part of
this company.

14. What are your goals? These may be both long and short-term
goals and might include learning various job-related skills, going
into management, improving your leadership skill capability, or
returning to school for a graduate degree.

15. What aspects of the profession interest you most? This might be
a big picture answer, such as the work being important to society,
or it might be more specific such as skill development.

16. How have you changed since starting college? You have matured
and you place importance on different things than you did before.
You might discuss how your goals and attitudes reflect this change.

17. Why are you interested in this job? Discuss what you are looking
for and how you believe this job satisfies those requirements. If
you believe the job will allow you to grow personally and professionally, discuss this, also.

18. Where do you expect to be in 10 years? Do you expect to be in
management? Do you expect to hold an advanced degree? Do
you expect to be in research? Try to provide some specific goals,
but also allow for flexibility so you do not give the impression
that you would not change to meet the company's goals for the
future.

In conclusion, provide specific examples when answering questions.
Mention projects that you worked on in college, as well as special qualities
that you possess. If you had leadership responsibilities in a part-time job
or in a group project for one of your classes, bring this up and emphasize
your ability to get along with others. Consider how special coursework
and abilities will enable you to contribute to the company. Stress your

accomplishments. Make sure you give no wild answers; conservative, positive, short answers are best.

11.2.2 Your Questions

To demonstrate your maturity and to show your interest in the company, ask the interviewer questions. This is also an opportunity to show that you have researched the company. If you ask mature questions, the interviewer will be favorably impressed. The questions should highlight your skills and knowledge, especially when they may have otherwise been overlooked by the interviewer's questions. It will be to your advantage to write out some specific questions prior to going to the interview and imagine the type of response that the interviewer may give. If you can also develop some pertinent follow-up questions to the interviewer's possible responses, then the interviewer will be especially impressed.

Some examples of questions that you might ask the interviewer are:

1. What computer system does your company use?
2. What training does your company provide?
3. Does the company reimburse the employee for advanced-degree educational expenses?
4. Do you encourage participation in professional society activities?
5. What are typical projects that I may be working on?
6. What will a typical working day be like?
7. Will there be any interaction with other parts of the company?
8. Will I be working as part of a team or will my work be more independent?
9. Will out-of-town travel be required?
10. At what point will I be given some management responsibility?

You should ask questions that deal directly with what you have learned about the company (this research will be discussed later) and about skills that you will develop as part of your job. Keep in mind that companies have a responsibility to help you grow professionally, and the responses to your questions should help you evaluate their interest in this responsibility.

Certain questions should be avoided as the interviewer will view them negatively. These include questions regarding vacation or sick days, expected starting time, and salary. After all, they are seeking a professional employee, not an hourly wage earner.

Asking the interviewer questions, particularly when he or she asks you if you have questions, will show that you are interested and enthusiastic about both the job and the company. It will also suggest that you possess a reasonable level of maturity and self-confidence. In some cases, the interviewer may not be very communicative, so your questions may actually put him or her at ease.

11.2.3 Pre-interview Research

Before going to the interview, you should research three items: the company, your values, and your monetary worth. Researching the company is important for the initial interview. When you are called in for a second interview, a more detailed research of the company and knowledge of your monetary value are necessary. Understanding your values will help you select a company where you will be happy.

Researching the Company. Researching a company prior to an interview will enable you to ask thoughtful questions and respond more favorably to the interviewer's questions. The recruiter will be impressed that you have some knowledge of the company.

For an initial interview, the research will, at a minimum, involve a trip to the career or placement office on your campus. Frequently, there is considerable information in brochures and pamphlets on the various companies. Look at the pictures in the brochures and note the environment of the company, but keep in mind that the brochure is essentially an advertisement and its intention is to give you a very positive first impression of the company. From the text in the brochure, get an idea of the company's philosophy, type of work, and products or services.

If you are called for a second interview, research the company in greater depth. Start by asking your professors about the company. Often, they will have had some experience, directly or indirectly, with the firm. Your professors and placement office may also be able to give you names of former students now working for the firm. These employees are probably the best source of information and are usually happy to discuss their job and company with students.

For other sources of information, talk with your librarian and visit the placement office. A librarian can direct you to trade journals, economic newsletters, and industry surveys, which can be valuable sources of information. The placement office may also have copies of annual reports for firms that hold interviews on campus.

When researching a company, have an outline of information you wish to obtain and take notes on each of the items. Don't expect to be able to recall everything you learn about a company, especially if you are interviewing with more than one. The following list of research items will assist you in assessing the company:

1. Products and services provided.
2. General information about the industry (growth/decline, government regulations, etc.).
3. Diversity and geographic location of company offices.

4. Financial situation.
5. Management philosophy and hiring policies.
6. Future direction.
7. Environment (i.e., supervision, dress, etc.).

Not only will this information be helpful in your interview, but it will also help you decide which company is best for you.

Researching Your Values. Beginning your professional life in the "real world" is a significant milestone. If you look back to your graduation from high school, you can probably recall that you had many concerns about which college to attend. You may have wanted a small school where you would not just be a number. You may have wanted to attend a college in a rural environment so you would not be near the congestion of an urban environment. You may have wanted to attend a college several hundred miles from home so that you could be somewhat independent of your parents.

These considerations were a reflection of what you valued at that time. But you've probably changed since high school and you need to consider the type of employment that will complement your current values. An honest assessment of what you value will enable you to select the position that will bring you the most satisfaction and stability.

Some people are interested in gaining management responsibilities after the first few years of working. Others prefer sticking to the technical work environment. Some even have the goal of starting their own company within five years of graduation, which would require considerable management responsibility. Where on the management/responsibility scale do your aspirations lie? Try to formulate questions to ask the interviewer so you will be able to assess the potential for a company to meet your managerial aspirations.

Do you value an advanced degree? You may have considered going to graduate school, but decided to put that off for a year or two. If an advanced degree is in your plans, possibly on a part-time basis, then you will want to choose a company that will support this goal. Additionally, you may consider the proximity to a local university when making the final job selection.

Do you value the open space of a rural environment or the opportunities of a major urban center? Believe it or not, when you select a company you must give equal weight to your personal and professional growth. If you take a job in a large city because the job offers many exciting professional opportunities, you may still be dissatisfied because you have difficulty coping with the congestion of the urban environment. If you value the pleasures of skiing, then you may want to take a job that is located where you could easily get away for weekend skiing trips.

Do you value travel? Some companies routinely manage projects that are not in close proximity to the office. Some companies have offices throughout the region or even nationally, and they want employees who are willing to travel to these locations. In some cases, travel is absolutely necessary. However, some people do not like to travel because of family responsibilities or the disorderliness associated with traveling. It is important to assess your attitude toward on-the-job travel and consider it when selecting an employer.

Management responsibility, educational advancement, locational preferences, and travel are a few of the value-based considerations that you should consider when you make your value assessment. These are important when formulating questions for the interview and making the final decision. While you may consider some of these value preferences informally, it is best to systematically assess your value preferences, and it is better to do this before an interview and well before you are under the pressures of accepting a job.

Researching Your Monetary Worth. During the initial interview, it is best to avoid any discussion of money, unless the interviewer specifically addresses the subject. However, it is important for you to be aware of your monetary worth. Salary discussions take place during the second interview or when a job offer is made. If you ask for too much, you may price yourself out of the market. If the value is too low, you may start your new position at a lower salary than some other employees. Since raises are often based on starting salaries, it may be quite some time before you catch up.

Knowing your monetary worth will help you get a competitive salary. If you believe that a salary offer is too low, don't be afraid to bargain with the employer. If a higher offer is made over the telephone, make sure you get the offer in writing before accepting the job and notifying other firms that you have accepted another offer.

There are several ways to determine your monetary worth. The best way is to ask recent graduates in your field what their starting salaries were. If you don't know anyone, ask your advisor or other professors. Some professors make formal or informal salary surveys of graduating seniors. This will give you a range of salaries. If you have special characteristics or abilities, you may ask for a starting salary at the upper end of the range. If you have a low GPA, no practical experience, and no special skills, then you may have to settle for a starting salary at the low end of the scale. Other sources of information on salaries are professional journals, professional societies, classified ads in newspapers, or job notices posted on bulletin boards near faculty offices.

In addition to the actual salary, the company will offer fringe benefits such as graduate school tuition. These benefits are an important part of overall income. Determine which are important to you and try to find out about these benefits. This topic should be discussed with the company after the initial interview, unless the company has a benefit package that is especially attractive; then it is fair to discuss it at the initial interview.

11.2.4   Appearance

You need to go shopping! It is a rare student who has appropriate clothing for a job interview. Many students cringe at the thought of purchasing a new suit or dress and shoes; students are frequently low on cash. However, bear in mind that $300 spent on new clothing and shoes is a small investment for a $20,000 job.

Remember the old saying, "First impressions are lasting impressions." Your appearance gives the interviewer his or her first impression of you. Make it count! Both clothing and hair should be conservative, not casual. Men should wear a dark colored suit and white shirt. Women may wear either a dress or a dark-colored suit and white blouse. Hair for both men and women should be clean, trimmed, and neat. Women should avoid wearing very long hair to an interview; if your hair is long, put it up. Women should keep makeup and jewelry to a minimum. Before purchasing new clothes, you may want to research the topic of dress. A number of "dress for success" books on the market, for both men and women, are excellent sources of information. You can look through clothing catalogs that include business fashion and assess the styles and colors of the most successful–looking clothes. You should note the dress of businesspersons who make presentations to student organizations. You could visit the campus placement office a month before you interview and note the dress of interviewers who are currently holding the interviews. You should probably not use your professors as role models for dressing.

11.2.5   Relaxation

Job interviewing can be stressful. However, if you are well prepared, the stress will be minimal, although a little nervousness is certainly to be expected. Preinterview jitters can actually be helpful to you. A controllable level of excitement will help keep your mind clear, keep you on your toes, and help you exhibit enthusiasm at the interview.

The interview process will be more pleasant if you are relaxed on the day of the interview. The best way to relax is to be completely ready for your interview at least two days ahead of time. That way there will be no pressure to get your new clothes, get your hair cut, or to finish researching the company. In addition, you should get plenty of fresh air and exercise. The day before the interview, go on a long walk or run, go swimming, play a game of tennis, or whatever other form of relaxing exercise you enjoy. That evening get a good night's sleep. You'll wake up refreshed, calm, and ready to tackle the interview.

11.3   THE DAY OF THE INTERVIEW

The day is here. Get yourself ready and allow enough time to get to the interview five or ten minutes early. Nothing will count against you like being late; it suggests immaturity and shows disrespect and a lack of enthusiasm and concern for the interviewer.

You should take a few items with you to the interview. Be sure to bring along an extra copy or two of your resume, a watch, and a pad and pencil so that you can jot down information that you might otherwise forget.

When you greet the interviewer, shake hands with a firm handshake (but don't crush his or her hand!) and look the interviewer in the eye. Eye contact is an important part of exhibiting confidence and a desire to speak with this person. During the interview, relax and smile. Take time to answer the questions carefully. Speak clearly and calmly and avoid talking about money. Let the interviewer lead the conversation, but participate enthusiastically in the discussion. The recruiter usually has a clear idea of the direction he or she wants the interview to take, but you want to make sure that the interviewer gets a complete understanding of your strong points. While it may be best for the interviewer to lead the interview, you should not be passive. Remember, while the interviewer is getting to know you and your capabilities, it is also a time when you have a right to learn about the company.

11.4   INTERVIEW FOLLOW-UP

The day after the interview, be sure to follow up with a letter thanking the interviewer for the time he or she has spent with you and for providing you with an opportunity to become more familiar with the company. The letter need not be more than one paragraph. The purpose of the letter is to demonstrate business courtesy and to get your name in front of the interviewer's eyes one more time. It is very important to spell the interviewer's name correctly in the follow-up letter; therefore obtain a business card from the interviewer. Even if the interviewer is not the personnel manager, send the letter to the interviewer since she or he will have significant input to the hiring decision.

If you do not hear from the employer in two or three weeks, send another letter stating your continuing interest in a position with the company. If your resume changes, be sure to send an updated version.

11.5   EXERCISES

1. While general questions about your work experience and leadership skills are part of an interview, questions about your technical specialty will also be asked. Prepare a list of five technical questions related to your academic major.

2. Your grade point average may not be as high as you would like it to be and the interviewer may have made a statement that the firm usually hires only those who have at least a 3.5 GPA. In such a case, you may believe that your GPA is a negative aspect of your qualifications. Prepare a statement that would put your less-than-desirable GPA in a positive tone.

3. Why would an interviewer be interested in your leadership experience? What activities might a student have that would improve one's leadership skills?

4. To help organize your thoughts on the importance of your profession to society, prepare a one-page summary on the subject.

5. Prepare a one-page statement on the questions, where do you expect to be professionally in 5 years? in 10 years?

6. Assume that you will ask what training the company provides. Prepare a list of possible responses that the interviewer may give and prepare one follow-up question that you might ask for each response.

7. Go to the campus job placement center and select five brochures for companies that hire students in your technical specialty. Analyze each and assess the type of work performed by the company, the products and services provided by the company, the extent to which the company has state-of-art equipment, and the work environment. Make a list of questions you might ask interviewers from each of these companies.

8. Contact two students you knew before their recent graduation who are currently employed in the technical specialty of interest to you. Ask them about important aspects of their current employment that they did not consider in interviewing and selecting a company.

9. Create a list of criteria that you will use in deciding which offer to accept. Try to rank these in importance and develop a scheme for measuring the extent to which a firm meets your desires.

10. Conduct a salary survey and develop a range of starting salaries. Contact recent graduates and professors and peruse classified advertisements to develop a histogram of salaries (make necessary adjustments for inflation). Decide upon a salary that you believe you should receive. Consider the reduction in salary that you would accept for a job that provides other opportunities.

11. Look through a mail-order catalogue of professional clothing. For women, what accessories are worn by the most professionally dressed models? For men, what are the colors and patterns of the most professional-looking ties? For both women and men, what colors and styles of shoes are worn by the most profession-

ally dressed models? Make a list of the prices for a complete outfit that you would want to wear to an interview.

12. Make a list of topics that could be discussed in a follow-up letter and draft a follow-up letter to an imaginary interviewer. Evaluate your letter on the basis of its positive tone and the degree to which it emphasizes the contributions that you could make to the firm.

13. Read through the list of interview questions and think about how you would answer each. Give the list to a friend and have the friend ask you the questions so that you can practice verbalizing your answer.

14. Write a follow-up letter thanking someone for an interview. After the letter is perfect, place it in your "jobs" file so that when you need it, you'll already know exactly what to write.

15. Pretend a friend is the interviewer and practice greeting the interviewer. Don't forget a firm handshake, eye contact, and a smile.

CHAPTER 12 / **NONVERBAL COMMUNICATION**

DO

- offer a firm handshake
- maintain good eye contact
- keep posture alert, but relaxed
- smile, when appropriate
- nod head occasionally
- lean forward slightly in seat
- keep arms unfolded
- wear subtle colors and accessories
- wear comfortable clothing and shoes
- keep hair, beards, and moustaches clean and neatly trimmed

DON'T

- jingle pocket change
- look everywhere but at the person with whom you are conversing
- cross arms
- wring hands
- offer a weak handshake
- frown or give an unfocused stare
- overuse gestures
- dress casually
- wear tight clothes
- wear clothing with bold patterns or loud colors

12.1 INTRODUCTION

Consider the following situations:

1. You believe that your test was unfairly graded, so you go to the instructor's office to ask her to regrade your test. While you are showing the instructor where you think the test was misgraded, she appears to be paying attention but folds her arms across her

chest. Do you believe the instructor has a positive view about your arguments?

2. Imagine that you work for a company that wants to subcontract work to other companies. During an interview with one potential subcontractor, he almost always looks toward the floor. What impressions might you develop about this individual?

These examples show that messages can be transmitted without words. When we speak with another person or to a group of people, we communicate our message verbally but also nonverbally. Facial expressions, posture, hand movements, and eye contact are all part of the nonverbal communication that may either support or negate the words. Most of a message is sent via a form of communication called body language or kinesics.

We both send and receive messages via body language. Therefore, in dealing with superiors or employees, it is very important both to watch for nonverbal signals *and* to be aware of the nonverbal signals that you are transmitting.

There are many benefits to translating body language. You can send messages such as, I am confident, I am a mature individual, or I know what I am talking about. If you are not aware of the meaning of body gestures and signals, you may be sending messages such as I am nervous, I dislike talking to you about this topic, or I am not confident of my position. Understanding the signals sent by others can enable you to put a nervous person at ease or recognize when someone is not being totally truthful with you.

This chapter will help you become aware of the meaning of various body gestures and signals and their importance in professional communication.

12.2   THE SIGNALS

Although there are many different gestures, we will discuss only those that are most important in business meetings, speeches, and job interviews. These include eye contact, handshakes, facial expressions, arm and hand movements, and posture. Body language in social situations will not be discussed, but there are many popular books available. You should consider reading one of these books so that you will be aware of the broader role of kinesics in everyday activities.

12.2.1   Eye Contact

Eye contact is one the most important forms of nonverbal communication. Good eye contact communicates confidence, attentiveness, and interest in communicating with the other person or the audience. Poor eye contact, on the other hand, is a sign of a lack of confidence, a low interest level, and anxiety.

At an employment interview, good eye contact with the interviewer will be regarded as a positive communication signal. By making eye contact with the interviewer, you will leave the impression that you would make a confident employee, one who would be a good representative of the company. Positive eye contact by the interviewer will suggest that he or she would look favorably upon you being hired by the company. At a business meeting, others will be more interested in what you have to say if you are making eye contact with them as you speak. The eye contact tells the business associate that you have important information or ideas that will help them; it tells them that you want their thoughts on the subject. When someone else is speaking, he or she will be encouraged by your attentiveness if you are making eye contact. It tells him or her that you are genuinely interested in them and their ideas.

Maintaining eye contact is also necessary to detect the other person's body signals. If you are not looking at the person, it will be difficult to catch their subtle gestures. You will not be able to judge whether or not they are confident in their knowledge or position. Not knowing this can lead to your inefficiency.

12.2.2 Handshake

A handshake is generally the initial form of contact between business associates or an interviewer and interviewee. It is important to have a handshake that will make a positive impression. A firm handshake, particularly in combination with eye contact, suggests energy, enthusiasm, and confidence. A wimpish handshake suggests subservience, a lack of confidence, and immaturity.

A good handshake takes practice. Practice your handshake with a friend until you have a pleasant, firm, confident handshake. No one likes a mushy handshake. On the other hand, a crushing handshake is likewise unpleasant.

12.2.3 Facial Expressions

What facial expression could be more pleasant than a smile? When used at the appropriate time, a smile is a very positive form of nonverbal communication that signals that you are warm, friendly, and in control. Since the smile is such a powerful expression, why not use it? There are times when smiling is inappropriate, such as in the middle of a serious discussion. A smile then can transmit a message of nervousness. It would be appropriate to smile when meeting someone at the beginning of an interview. It would be inappropriate to smile when the interviewer asks you about your work experience or your academic background in statistics.

Facial expressions are also used to communicate truthfulness, interest and understanding. When you are talking with someone, he or she will be more inclined to believe your words if they are supported by your facial expressions. For example, if you tell one of your professors that you did not complete a homework assignment because you were sick, but you smile profusely while giving your excuse, the professor is unlikely to believe you; your facial expressions do not support your words.

A blank facial expression is somewhat the opposite of a smile. It implies boredom with the subject or person speaking. A blank expression is common when the individual's mind is on something other than the speaker. This lack of variation in expression is often a signal that the expressionless person wishes to end the conversation. The receiver of the blank expression usually feels rejected. So it is important to avoid giving someone the feeling that you are not interested in them. An occasional nod of the head will imply understanding and interest; it will make the other person believe that you are genuinely interested in their ideas.

Other movements of the head send messages. For example, leaning your head on a hand indicates boredom and disinterest. A hand across the face signifies doubt.

12.2.4 Arm and Hand Movements

Arm and hand movements may also send positive or negative signals. Crossed arms usually indicate that the listener is finished with the subject at hand, but it may also indicate fear, timidity, and restraint. For example, when the instructor crossed her arms during the discussion with the student who was asking her to regrade the test, it indicated that she was not receptive to the idea. When a speaker crosses his or her arms, it usually suggests a fear of the audience or their reaction to his or her ideas. Since fear, timidity, or restraint are not positive attributes, it is best to avoid crossing one's arms during a business meeting of any type. It suggests that you are closing off the other person from the discussion, and no one wants to feel excluded. Conversely, open arms indicate a receptive attitude. It is a positive form of body language, especially when making a speech.

When a television crew is interviewing someone who has lost a family member in a tragic accident, the camera often focuses on the individual's hands. Wringing hands indicate despair and grief. The act can be just as telling as tears. Your hands may also send signals in a professional setting. Wringing hands is a sign of nervousness and should always be avoided since nervousness signals a lack of confidence. Placing your hands on your hips is an authoritative signal. Hands palm down on a desk shows an eagerness to get down to business. Open, outstretched hands are indicative of sincerity and openness; they welcome the listener into the discussion. Again, signals transmitted through hand movements can be positive or negative. You should be aware of signals you are sending with your hand movements. Practice the positive indicators and avoid using the negative hand motions.

12.2.5 Posture

The way you sit or stand can have a positive or negative impact on the person or persons receiving your message. An alert but relaxed posture is best. Slumped shoulders and a bowed head may be viewed negatively, while erect posture displays self-confidence. Posture should be practiced so that good posture is easy and comfortable.

12.3   THE SITUATIONS

12.3.1   Interviews

Preparation for a job interview should include gaining an awareness of the impact of nonverbal communication. Eye contact, handshake, and posture should be practiced. Interviewers are often seasoned veterans of the interviewing process and are aware of nonverbal signals. Some are even trained in nonverbal communication so that they may both present the proper signals and quickly interpret yours. Therefore, it behooves you to be familiar with nonverbal communication. The proper use and awareness of nonverbal signals may be the edge you need over other applicants to get the job.

Remember that first impressions are lasting impressions. Therefore, greet the interviewer with a smile, good eye contact, and a firm handshake. It will get the interview off to a very positive start. During the interview, it is important for you to be aware of the interviewer's signals as well as your own. For example, if while you're talking, the interviewer crosses his or her arms and gives you an unfocused look, he or she has most likely tired of the current subject. At that point, you should try to turn the interview into a more positive discussion; specifically, get the interviewer talking about his or her company.

During an interview, show interest and enthusiasm to the interviewer. Lean forward in your seat and occasionally nod your head to indicate interest. Conversely, don't wiggle in your seat, cross your arms, and fail to make eye contact. During the interview, you will choose your words carefully; just remember, it is just as important to choose your body language carefully.

12.3.2   Speeches

When you have to make an oral presentation, you should practice until the delivery is smooth. During these practice sessions, it is equally important to practice the nonverbal as well as the verbal communication. Practicing control of your body signals will help you appear to be a polished, confident speaker. It will be a significant factor in determining your effectiveness.

Before beginning your speech, address the group with a smile. This will put them in a positive frame of mind and make them want to listen to you. Be sure that any other smiles used during the talk are appropriate and not forced. You do not want a wayward smile to suggest nervousness.

During a speech it is essential to maintain eye contact with the audience. The audience will be more attentive if they know you are watching them. Stand erect without crossing your arms, keep your legs slightly and comfortably apart, and hold your head high. These body movements strongly signal confidence. If you have visual aids, don't be afraid to point to the figure as you speak. This motion also demonstrates confidence.

12.3.3 Business Meetings

You're meeting with one of your superiors. You're nervous. You want to wring your hands, but you know that would be a sure sign of nervousness. Instead, you concentrate on a firm handshake, eye contact, and a warm smile. While the boss is talking, you listen attentively, maintain eye contact, nod your head occasionally in understanding, and lean slightly forward in your seat to show your attentiveness and interest. You should think about this prior to the meeting, as it will probably not occur spontaneously.

When it's your turn to talk, keep your arms uncrossed, stand or sit in an alert, relaxed manner, and maintain eye contact. Your superior will certainly believe that you are interested, enthusiastic, and confident.

If you are meeting with a group, be sure to watch for signs of acceptance from the others in the room. Make sure everyone is relaxed and leaning forward in their seats with anticipation and approval. Interpret signals from the audience and adjust your presentation in a way that will elicit more positive nonverbal signals from the audience.

12.4 COMMUNICATION THROUGH DRESS

When the president greets a foreign diplomat at the White House, what does he wear? Probably a dark suit with a white shirt is the standard. Do you think the foreign diplomat would be insulted if the president appeared in shorts and a T-shirt?

Imagine the president of General Motors walking into the conference room for a meeting with the Board of Directors. What do you imagine he is wearing? Do you believe that his attire will influence the level of support the GM president will receive from the board?

Do clothes influence your assessment of an individual? If a professor arrives to your class on the first day of classes dressed very casually, does it lower your expectations for the class? Conversely, if the professor is dressed very professionally, does it leave you with a more positive attitude?

People dress in a variety of ways for a variety of reasons: to give certain impressions, for comfort, or for practical reasons. A car mechanic wears sturdy overalls to keep grease, heat, and sharp objects away from his or her body. A popular stage performer wears glamorous or outrageous clothing to give the audience something fun to look at. A person going on a 10-hour driving trip might wear a sweat suit to be comfortable. A chemistry student going to his or her first job interview will wear a suit to impress the interviewer with his or her seriousness about the job.

Dress and other aspects of appearance will have a significant effect on your professional growth. Dress and appearance communicate your attitude and self-image. And, most importantly, they effect the feelings others have toward you. The purpose of this chapter is to discuss the role of appearance and clothes and the basic expectations of professionals in terms of dress and appearance.

12.4.1 The Effect of Appearance

Your clothing, accessories, and hairstyle are primary factors in the impression you make on people. During a first meeting, such as an interview, this first impression is very important; it is the interviewer's first feeling about whether he or she might be interested in hiring you. Your appearance will influence people's assessment of your intelligence, maturity, and potential for success. We know that a black suit, black shirt, and white tie are the gangster look and that such dress leads to the assumption that the wearer has the value system of a gangster. So your appearance influences other's impressions of your value system.

Your appearance is, of course, not the only thing that matters, but it is an important part of your complete package. A singer can come on a stage in beautiful or outrageous clothing but must also be able to sing! Pleasing dress, however, makes the show more entertaining. As an employee or job applicant, you need to dress appropriately so that the co-workers, superiors, or interviewers will respond favorably to you.

Your dress effects you as well. Appropriate dress will help you feel more confident and secure. Improperly dressed, your message is, "I don't look good, therefore I don't feel good, and thus I am not giving my best."

12.4.2 Clothing

Clothing is to the professional as packaging is to consumer goods and, as any marketing major will tell you, packaging is important to the success of a consumer product. Powdered soap in boxes of detergent is essentially the same across brands. Yet some brands get a much larger market share and others fail. Studies have clearly demonstrated that packaging makes a significant difference, and color is an important aspect of packaging. In addition to color, fabric type is important in marketing certain products such as furniture. Fit is a third important marketing parameter. For example, luggage that fits under an airline seat as well as in overhead luggage compartments is a very marketable consumer good. These parameters (color, fabric, and fit), as well as cost, are important aspects of clothing that influence your ability to sell yourself.

Colors of clothes appropriate for social functions may not be appropriate for business attire. This is due, in part, to different objectives in social and business activities. Although colors can make clothing fun, certain colors should be avoided in business attire. Those colors include pastels and bright or loud colors such as purple, red, and yellow. Pastels may reflect a too casual attitude and bright colors create a fun or party image.

Colors appropriate for men's business suits are subtle, muted colors such as medium or dark gray or navy blue. Men's ties may be maroon, blue, or other dark, subtle colors. These colors suggest that the wearer is a responsible person. While acceptance of these colors is largely cultural, acceptance has been stable over a long period of time. Brown is generally not a good color for men's suits. In addition to those colors, women's suits may be medium blue, forest green, or maroon. Shirts and blouses should be white or light blue. Other colors may also be appropriate, depending on the accepted level of formality in the profession, but the color should be pale. Women may also wear subtly colored dresses.

The best fabric for men's and women's suits is medium-weight wool which can be worn eight or nine months out of the year and looks good for a long time. Wool blends may actually be more desirable since they do not wrinkle as easily. New fabrics made entirely of rayon, but which look and feel like wool are also desirable for look, comfort, and care. For the warm weather, another suit of a lightweight wool blend may be necessary.

Shirts and blouses should be cotton or a cotton blend, accessorized with silk scarves for women or silk neckties for men.

Your clothes must fit well. Make sure your suit is well tailored and fitted, particularly in the shoulders. Men's slacks should be hemmed to just above the heel; pants should never drag on the ground. Skirt or dress hems should be at or just below the knee. Low-cut necklines on blouses or dresses are not acceptable.

Yes, business clothes cost a lot. Not only will you need a suit for an interview, but you will likely need another suit or two for your new job. If your budget is tight, as it probably is, try to plan ahead; rather than waiting until you need a suit, purchase your suit ahead of time. This way, you can get a reduced price by shopping at stores during sales or at outlet stores. You can also minimize expenditures for business clothes by avoiding faddish clothing and purchasing instead traditional, conservative clothing that will not be out of style in a year to two. Buying suits of medium-weight fabrics is economical because they can be worn most of the year.

12.4.3   Hair and Makeup

When you meet someone, you look first at their face. For this reason, women should be careful that their makeup is discrete. It should not make a statement, but subtly accentuate or enhance facial features. No one should look at your face and see makeup. Strong perfume should be avoided for daytime business wear.

Since most faces are surrounded by hair, it makes sense that the hair should also have a subtle accentuating look. Hair for both men and women should be clean, well-groomed, and short. Women with longer hair should make sure the hair is well away from the face. Men with beards may want to consider shaving them off prior to the interview. However, if you were hired with a beard, you should be able to keep it. Make sure that beards and moustaches are neatly trimmed close to the face.

12.4.4   Accessories

Accessories include shoes, jewelry, belts, scarves, hosiery, and so on. Shoes should be in good shape, polished, and clean. For men, black wing tips or other conservative shoes are best; loafers are not appropriate for business dress. For women, dark-colored pumps with heels two inches or less are appropriate.

Hosiery should be in good repair. Men should wear dark-colored socks that match the suit. Women should wear dark or neutral hosiery. It is also advisable for women to carry an extra pair of hosiery in their purse.

Silk scarves, leather belts, and jewelry can be used to dress up women's suits and dresses. Again, be sure the accessories are subtly colored. Jewelry for women should consist of a gold watch and simple pearl or gold earrings.

Men may also want to wear jewelry, but should wear only a gold watch or a wedding or scholastic ring. Please, no earrings!

Most men and women in any business will need a briefcase. The briefcase should be a solid color and preferably leather. Women will probably also wish to carry a purse. The purse should be small, tasteful, and preferably leather.

12.4.5   Recommendations

For any situation, it is best not to be under– or overdressed. The tips given here should be helpful in deciding how to dress for interviews and at work. The most important word in business dress is conservative. Conservative dress for men includes a two-piece suit, a silk tie with no wild designs, and dark socks. Women should wear a dress or a suit, never slacks. A job will occasionally be casual enough to permit women to wear slacks or men to wear sports jackets rather than a suit, but it is best to start out in a dress or suit.

Casual dress is never appropriate for an interview. Remember, this is your first impression on the interviewer, make it count. Avoid sports jackets, faddish clothing, and bold colors or patterns. Muted tones such as gray and navy blue are best for suits. The shirt or blouse should be white.

Suits may seem uncomfortable at first. Try it on a few days ahead of the interview; wear it around the house to get used to it. You should not fidget in your suit at the interview. You want to be comfortable and confident.

Different companies have different dress standards. Some companies are relatively casual in their dress standards; others are more formal. An easy way to find out what sort of clothing the employees wear at your new company before you start is to ask someone who works there or to watch people leaving the building at quitting time.

You are probably interested in "getting ahead" at work. In this case, it is advisable to dress appropriately for the position you would like to hold next. You may emulate the dress of your immediate supervisor.

12.5 EXERCISES

1. Compile a list of about 10 books that deal, at least in part, with the subject of body language.
2. Position yourself where you can observe two people talking, such as an after-class discussion between a student and a professor, but where you cannot hear the discussion. Observe the two people and try to assess the nature of the conversation entirely from the body language that the two exhibit. If possible, talk to one of the participants after their discussion is complete and evaluate the accuracy of your assessment.
3. Attend a seminar or student organization meeting where an outside speaker is featured. Throughout his or her presentation, assess his or her nonverbal signals. Notice whether the speaker made continual eye contact with the audience, crossed his or her arms, stood facing the audience with good posture, and opened the presentation with positive facial expressions. After the presentation, talk with other people in the audience and assess their views on the speaker's effectiveness. Is there a correlation between the speaker's effectiveness and the assessment you made based on the speaker's body signals?
4. When you visit your professors to get help with an assignment, practice making eye contact as you speak with them. The more you practice this, the easier and more natural it becomes.

5. With a group of friends, shake hands with each one of them. What does each handshake tell you about the person? Which handshake is the most confident? Practice your handshake with your friends until it is firm and confident.

6. Go to a conference in your field and note the mannerisms of the speakers. What body movements tell you the speaker is: nervous? unconfident? bored? Which speakers seem confident? How is this confidence displayed by their body language?

7. During a lunch hour, walk through the business district of your local city or town. Observe the dress of the executives. Assess the attire of the more distinguished looking men and women and decide what sets them apart from the others.

8. Using a mail-order catalogue for business clothes, try to characterize the qualities of the more expensive suits. Evaluate suit-and-tie combinations of men's clothing or suit-and-scarf combinations for women's clothing.

9. Visit the career center at your college and observe the attire of the interviewers. Make a list of clothes that you need to purchase in order to have a similar appearance.

10. Note the dress of the following people: business executives, professors and sales people. What is the difference in the way people in these three types of occupations dress? What message is conveyed by their attire? What message should each group be conveying?

11. Imagine the following situations: you are notified by one of your professors at 10 AM that a certain recruiter is on campus. Being certain that you would be interested, he has scheduled an interview for you for 1:00 PM. You dash home and throw on some clothes. In the first case, you don't feel you have adequate time to dress up and so you choose to wear slacks and a button-up shirt or blouse. In the second case, you take the time to put on a suit and fix your hair (you have previously thought about how to do this). In both cases, you are a bit frenzied and rushed and arrive at the interview without having given much thought to what questions you will ask or how you will answer the interviewer's questions. Describe how you think the two manners of dress will affect your mood and the impression you make during the interview.

INDEX